W0232124

Encyclopaedia of
Mathematical Sciences
Volume 17

Editor-in-Chief: R.V. Gamkrelidze

A.V. Arkhangel'skiĭ   L.S. Pontryagin (Eds.)

General Topology I

Basic Concepts and Constructions
Dimension Theory

With 15 Figures

Springer-Verlag
Berlin Heidelberg New York
London Paris Tokyo
Hong Kong Barcelona

Consulting Editors of the Series: N.M. Ostianu, L.S. Pontryagin
Scientific Editors of the Series:
A.A. Agrachev, Z.A. Izmailova, V.V. Nikulin, V.P. Sakharova
Scientific Adviser: M.I. Levshtein

Title of the Russian edition:
Itogi nauki i tekhniki, Sovremennye problemy matematiki,
Fundamental'nye napravleniya, Vol. 17, Obshchaya topologiya 1
Publisher VINITI, Moscow 1988

Mathematics Subject Classification (1980):
54-01, 54-02, 54F45

Library of Congress Cataloging-in-Publication Data
Obshchaiā topologiiā I. English. General topology I / A.V. Arkhangel'skii, L.S. Pontryagin, eds.
p. cm.—(Encyclopaedia of mathematical sciences; v. 17)
Translation of: "Obshchaiā topologiiā I," which is vol. 17 of the serial
"Itogi nauki i tekhniki. Seriiā Sovremennye problemy matematiki. Fundamental'nye napravleniiā."
Includes bibliographical references.
Contents: The basic concepts and constructions of general topology / A.V. Arkhangel'skii, V.V. Fedorchuk—
The fundamentals of dimension theory / V.V. Fedorchuk.
ISBN-13: 978-3-642-64767-3   e-ISBN-13: 978-3-642-61265-7
DOI:10.1007/978-3-642-61265-7
1. Topology. 2. Dimension theory (Topology) I. Arkhangel'skiĭ, A.V.
II. Pontriāgin, L.S. (Lev Semenovich), 1908–
III. Arkhangel'skiĭ, A.V. Osnovnye poniātiiā i konstrukt͡sii obshchei topologii. English. 1990.
IV. Fedorchuk, V.V. Osnovy teorii razmernosti. English. 1990.
V. Title. VI. Title: General topology 1. VII. Series.
QA611.02713   1990   514'.322—dc20   89-26209

This work is subject to copyright. All rights are reserved, whether the whole or part of the material in concerned,
specifically the rights of translation, reprinting, reuse of illustrations, recitation, broadcasting, reproduction on
microfilms or in other ways, and storage in data banks. Duplication of this publication or parts thereof is only
permitted under the provisions of the German Copyright Law of September 9, 1965, in its current version,
and a copyright fee must always be paid. Violations fall under the prosecution act of the German Copyright Law.
© Springer-Verlag Berlin Heidelberg 1990
Softcover reprint of the hardcover 1st edition 1990

Typesetting: Asco Trade Typesetting Ltd., Hong Kong
2141/3140-543210—Printed on acid-free paper

List of Editors, Contributors and Translators

Editor-in-Chief

R.V. Gamkrelidze, Academy of Sciences of the USSR, Steklov Mathematical Institute, ul. Vavilova 42, 117966 Moscow, Institute for Scientific Information (VINITI), Baltiiskaya ul. 14, 125219 Moscow, USSR

Consulting Editors

A.V. Arkhangel'skiĭ, Moscow State University, Department of Mathematics and Mechanics, Chair of General Topology and Geometry 119899 Moscow, USSR
L.S. Pontryagin, Steklov Mathematical Institute, ul. Vavilova 42, 117333 Moscow, USSR

Contributors

A.V. Arkhangel'skiĭ, V.V. Fedorchuk, Moscow State University, Department of Mathematics and Mechanics, Chair of General Topology and Geometry, 119899 Moscow, USSR

Translator

D.B. O'Shea, Department of Mathematics, Mount Holyoke College, South Hadley, MA 01075, USA

Contents

I. The Basic Concepts and Constructions of General Topology

A.V. Arkhangel'skiĭ, V.V. Fedorchuk

Translated from the Russian
by D. O'Shea

Contents

Introduction

General topology is the domain of mathematics devoted to the investigation of the concepts of continuity and passage to a limit at their natural level of generality. The most basic concepts of general topology, that of a topological space and a continuous map, were introduced by Hausdorff in 1914.

One of the central problems of topology is the determination and investigation of *topological invariants*; that is, properties of spaces which are preserved under homeomorphisms.

Topological invariants need not be numbers. Connectedness, compactness, and metrizability, for example, are non-numerical topological invariants. Dimensional invariants, on the other hand, are examples of numerical invariants which take integer values on specific topological spaces. Part II of this book is devoted to them. Topological invariants which take values in the cardinal numbers play an especially important role, providing the raw material for many useful computations. Weight, density, character, and Suslin number are invariants of this type.

Certain classes of topological spaces are defined in terms of topological invariants. Particularly important examples include the metrizable spaces, spaces with a countable base, compact spaces, Tikhonov spaces, Polish spaces, Čech-complete spaces and the symmetrizable spaces.

The *main* "internal" *tasks of general topology* are: 1) the comparison of different classes of topological spaces; 2) the study of the spaces in a given class and the categorical properties of the class as a whole; 3) the determination of new classes of topological spaces, the appropriateness of which resides in the pursuit of 1) and 2) and the needs of applications. In addressing 1), the method of jointly classifying spaces and maps is particularly important. The method is directed at establishing connections between different classes of topological spaces through continuous maps subordinate to certain simple constraints. Related to 2) is the question of determining which classes of spaces are closed with respect to the operations of taking products and passing to subspaces. The method of coverings plays an important role in investigating 1), 2) and 3).

This article in introductory: it introduces the basic concepts of general topology, sketches the most instructive examples and describes the most basic constructions of general topology. Although many important theorems are quoted, the presentation is very broad and makes no pretensions to depth. The landscape and methods of general topology are merely outlined.

Chapters 1–5 of this article were written by A.V. Arkhangel'skiĭ, and chapters 6–12 by V.V. Fedorchuk.

§1. Topological Spaces: The First Notions

1.1. The Definition of a Topology and a Topological Space

Definition 1. A *topology* on a set X is a collection $\mathcal{T}$ of subsets of X which includes the empty set $\varnothing$ and the whole set X and which is such that: a) the intersection of any finite number of elements of $\mathcal{T}$ belongs to $\mathcal{T}$ and b) the union of any set of elements of $\mathcal{T}$ belongs to $\mathcal{T}$.

Example 1. The collection $\{\varnothing, X\}$ consisting of just two elements, the empty set and the whole set X, is a topology on X. This topology is called the *antidiscrete topology* on X.

Example 2. The set Exp X of all subsets of a set X is also a topology on X. This topology is called the *discrete topology*.

Example 3. Let $X = \{0, 1\}$ be a two point set. The collection $\mathcal{T} = \{\varnothing, \{0\}, \{0, 1\}\}$ is a topology on X.

Thus, different topologies can be introduced on any set containing more than one point.

Definition 2. A set X together with a fixed topology $\mathcal{T}$ is called a *topological space*. The space is denoted by $\{X, \mathcal{T}\}$ or, more briefly, as X. The elements of the topology $\mathcal{T}$ are called *open sets* in X.

A point $x \in X$ is said to be *isolated* in a topological space X if the one point set $\{x\}$ is open (that is, $\{x\} \in \mathcal{T}$). A space X is said to be *discrete* if all its points are isolated. It is clear that X is a discrete space if and only if all its subsets are open (that is, $\mathcal{T} = \text{Exp } X$).

If $\mathcal{T}_1$ and $\mathcal{T}_2$ are topologies on the same set and if $\mathcal{T}_1 \subset \mathcal{T}_2$, then the topology $\mathcal{T}_1$ is said to be *weaker* than $\mathcal{T}_2$ and the topology $\mathcal{T}_2$ is said to be *stronger* than $\mathcal{T}_1$. Of course, the discrete topology on X is stronger than any topology on X and the antidiscrete topology is weaker than any other topology on X.

If $\{\mathcal{T}_\alpha : \alpha \in A\}$ is any collection of topologies on a set X, then the intersection $\mathcal{T} = \bigcap \{\mathcal{T}_\alpha : \alpha \in A\} = \{U \subset X : U \in \mathcal{T}_\alpha, \text{ for all } \alpha \in A\}$ is also a topology on X. This topology is weaker than all the topologies $\mathcal{T}_\alpha$ where $\alpha \in A$. In particular, if $\mathcal{T}_1$ and $\mathcal{T}_2$ are any two topologies on X, their intersection $\mathcal{T}_1 \cap \mathcal{T}_2 = \{U \subset X : U \in \mathcal{T}_1 \text{ and } U \in \mathcal{T}_2\}$ is a topology on X.

1.2. Subbases and Bases of a Topological Space.

Let $\mathscr{E}$ be a collection of subsets of a set X. Let $\mathscr{B}_\mathscr{E}$ be the smallest collection of subsets of X which contains all the elements of $\mathscr{E}$ together with the empty set and the whole set X and which is closed under finite intersection. The latter means that if $U_1, \ldots U_k \in \mathscr{B}_\mathscr{E}$, then $U_1 \cap \cdots \cap U_k \in \mathscr{B}_\mathscr{E}$. Let $\mathcal{T}_\mathscr{E}$ be the collection of all sets which are a union of some set of elements in $\mathscr{B}_\mathscr{E}$; that is, $\mathcal{T}_\mathscr{E} = \{\bigcup \gamma : \gamma \subset \mathscr{B}_\mathscr{E}\}$. It is easy to check that $\mathcal{T}_\mathscr{E}$ is a topology on X; in fact, $\mathcal{T}_\mathscr{E}$ is clearly the smallest topology on X containing $\mathscr{E}$. The collection $\mathscr{E}$ is called a *subbase of the topology* $\mathcal{T}_\mathscr{E}$ and we say that the topology $\mathcal{T}_\mathscr{E}$ is generated by the subbase $\mathscr{E}$.

Definition 3. A collection $\mathscr{B}$ of open subsets of a topological space X is called a *base* of the space (or of the topology) if each open set in X is a union of some elements of $\mathscr{B}$. Note that the whole topology $\mathscr{T}$ is a base of the space $(X, \mathscr{T})$.

It is easy to verify that a collection $\mathscr{B}$ of subsets of X is a base of the topological space $(X, \mathscr{T})$ if and only if all elements of $\mathscr{B}$ are open and, for every point $x \in X$ and each set $U \in \mathscr{T}$ for which $x \in U$, there exists $V \in \mathscr{B}$ such that $x \in V \subset U$.

Since $\mathscr{T} = \{\bigcup \gamma \colon \gamma \subset \mathscr{B}\}$, it follows that a collection $\mathscr{B}$ cannot be a base of two different topologies on the same set X; that is, the topology is uniquely determined by a base. On the other hand, a single topology can have many different bases (see the examples below). This circumstance lends the concept of a base much of its flexibility: one can choose a base with a variety of special properties. For example, a topology $\mathscr{T}$ might be uncountable, but have a countable base. This greatly facilitates the analysis of the structure of such spaces by allowing the use of arguments based on countable procedures. Other advantages associated with the consideration of bases accrue from the fact that some bases have an especially simple structure or have elements that admit a very concrete description.

The following simple, but very useful, assertion completely answers the question of when a collection of subsets of a set X is a base of a topology on X.

Proposition 1. *A collection $\mathscr{B}$ of subsets of a set X is a base of a topology on X if and only if $\bigcup \mathscr{B} = X$ and for any $U, V \in \mathscr{B}$ and any $x \in U \cap V$ there exists $W \in \mathscr{B}$ for which $x \in W \subset V \cap U$.*

In particular, a collection $\mathscr{B}$ of subsets of X which covers X and is closed under finite intersection is a base of the topology $\mathscr{T} = \{\bigcup \gamma \colon \gamma \subset \mathscr{B}\}$. For example, $\mathscr{B}_{\mathscr{E}}$ is a base of the topology $\mathscr{T}_{\mathscr{E}}$ generated by the subbase $\mathscr{E}$ (see the beginning of this section).

Example 4. The collection $\mathscr{B}$ of all intervals $(a, b) = \{x \in \mathbb{R} \colon a < x < b\}$ on the real line $\mathbb{R}$ covers $\mathbb{R}$ and is closed under finite intersection. Hence $\mathscr{B}$ is a base of a topology $\mathscr{T}$ on $\mathbb{R}$. This topology is called the *natural* or *usual topology* on $\mathbb{R}$. One can show that $U \in \mathscr{T}$ if and only if U is a union of pairwise disjoint intervals (see [A4]).

Example 5. Consider the collection $\mathscr{B}$ of those disks on the euclidean plane $\mathbb{R}^2$ which contain no points of the circles that bound them. The collection $\mathscr{B}$ is not closed under finite intersection, but does satisfy the conditions of Proposition 1. Hence $\mathscr{B}$ is a base of a topology on $\mathbb{R}^2$. This topology is called the *natural* or *usual topology on the plane*.

Example 6. Let $\mathscr{B}_s$ be the collection of all semi-intervals of the form $[a, b) = \{x \in \mathbb{R} \colon a \leqslant x < b\}, a, b \in \mathbb{R}, a < b$, on the line $\mathbb{R}$. It is clear that $\mathscr{B}_s$ is closed under finite intersection and that $\bigcup \mathscr{B}_s = \mathbb{R}$. Hence, $\mathscr{B}_s$ is a base of a topology $\mathscr{T}_s$ on $\mathbb{R}$. The topological space $(\mathbb{R}, \mathscr{T}_s)$ is called the *Sorgenfrey line*. Every interval (a, b) can be written as a union of elements of $\mathscr{B}_s$; hence, $\mathscr{T}_s$ is stronger than the usual topology on the line $\mathbb{R} \colon \mathscr{T} \subset \mathscr{T}_s$. Since $\mathscr{B}_s \subset \mathscr{T}_s \setminus \mathscr{T}$, we have $\mathscr{T}_s \neq \mathscr{T}$. In what follows, we shall often return to this example. For the moment, we mention only one of the remarkable properties of the Sorgenfrey line: the space $(\mathbb{R}, \mathcal{T}_s)$ does not have a countable base (the base $\mathcal{B}_s$ is clearly uncountable). The proof is based on the following very useful fact about bases.

Proposition 2. *Let $\mathcal{B}$ and $\mathcal{P}$ be two infinite bases of the topological space $(X, \mathcal{T})$. Then there exists a base $\mathcal{B}_0$ of $(X, \mathcal{T})$ such that $\mathcal{B}_0 \subset \mathcal{B}$ and $|\mathcal{B}_0| \leqslant |\mathcal{P}|$.*

◁ Let $t = (U, V)$ be an arbitrary pair of elements of the base $\mathcal{P}$. Let $\varphi(t)$ denote some element of $\mathcal{B}$ for which $V \subset W \subset U$, provided such an element exists. Otherwise, set $\varphi(t) = \varnothing$. The collection $\mathcal{B}_0 = \{\varphi(t): t \in \mathcal{P} \times \mathcal{P}, \varphi(t) \neq \varnothing\}$ is a base of the space X for which $\mathcal{B}_0 \subset \mathcal{B}$ and $|\mathcal{B}_0| \leqslant |\mathcal{P} \times \mathcal{P}| = |\mathcal{P}|$. ▷

◁ No countable subset γ of the base $\mathcal{B}_s$ of the Sorgenfrey line can be a base of the Sorgenfrey line. In fact, the set $A_\gamma = \{a: \exists b \in \mathbb{R}, [a, b) \in \gamma\}$ is countable. Choose a set $[c, d) \in \mathcal{B}_s$ where $c \in \mathbb{R} \setminus A_\gamma$ and $c < d$. It is clear that $[c, d)$ cannot be written as a union of elements in γ. It now follows from Proposition 2 that the Sorgenfrey line cannot have a countable base. ▷

1.3. Neighbourhoods of Points. Nearness of a Point to a Set and the Closure Operator

Definition 4. A set $O \subset X$ is called a *neighbourhood* of a point x in the topological space X if there exists an open subset U of X such that $x \in U \subset O$.

An open set is a neighbourhood of every point it contains. The intersection of two neighbourhoods of a point is again a neighbourhood of the point.

Intuitively, one should think of a neighbourhood of a point as the set of points "near" to the point, the neighbourhood itself serving as a measure of this closeness. One can also think of a neighbourhood of a point as a set of approximations to the point. The notion of a neighbourhood allows one to formally define what it means for a point to be near a set in a topological space. The nearness relation between sets and points is an absolute notion – intuitively, the assertion that a point is near to a set is best interpreted as saying that the point lies at distance zero from the set (although distance between points is not defined, and, in general, cannot even be adequately defined, in a topological space). Although absolute, the nearness relation between a point and a set is different from the containment relation; it is precisely the nearness relation upon which the mathematical constructions pertaining to the analysis of the concept of continuity are based.

Definition 5. Let $(X, \mathcal{T})$ be a topological space and let $A \subset X$ and $x \in X$. We say that the point x is *near* to (or at *distance zero* from) the set A and write $x\delta A$ (or $\delta(x, A) = 0$) if every neighbourhood U of x has nonempty intersection with A: $U \cap A \neq \varnothing$. If x is not near to A, then we write $x\bar{\delta}A$ (or $\delta(x, A) = 1$) and say that x is *far* from the set A. The relation δ between points and subsets of X is said to be *induced by the topology* $\mathcal{T}$; in order to emphasize this, we write

$\delta = \delta(\mathcal{T})$. The relation $\delta = \delta(\mathcal{T})$ possesses the following properties for every $x \in X, A \subset X, B \subset X$:

$1_\delta)$ $x \not\delta \varnothing$ (that is $\delta(x, \varnothing) = 1$);

$2_\delta)$ $x \in A \Rightarrow x\delta A$ (that is, $x \in A \Rightarrow \delta(x, A) = 0$);

$3_\delta)$ $x\delta(A \cup B) \Leftrightarrow x\delta A$ or $x\delta B$ (that is, $\delta(x, A \cup B) = 0 \Leftrightarrow \delta(x, A) = 0$ or $\delta(x, B) = 0$);

$4_\delta)$ If $x\delta B$ and $y\delta A$ for all $y \in B$, then $x\delta A$ (that is, if $\delta(x, B) = 0$ and $\delta(y, A) = 0$ for all $y \in B$, then $\delta(x, A) = 0$; this is a weak form of the triangle axiom!).

The absolute nature of the nearness relation between a point and a set is manifest in the main part of condition $3_\delta)$: if a point is far from the sets A and B then it is also far from their union; and in condition $4_\delta)$ which expresses the transitivity of the nearness relation. More precisely, $4_\delta)$ is analogous to the arithmetical identity $0 + 0 = 0$.

The set of all points near to a subset $A \subset X$ is the called the *closure* $\bar{A}$ of A in the topological space X. Thus, $\bar{A} = \{x \in X : x\delta A\}$. The map $A \to \bar{A}$ is called the *closure operator* in the topological space X. The closure operator possesses the following properties:

$1_c)$ $\bar{\varnothing} = \varnothing$.

$2_c)$ $A \subset \bar{A}$ for all $A \subset X$.

$3_c)$ $\overline{A \cup B} = \bar{A} \cup \bar{B}$.

$4_c)$ $\bar{\bar{A}} = \bar{A}$ for all $A \subset X$.

A set which contains all points near to it is said to be *closed*. Thus, the closed sets are simplest in the sense that the nearness relation is trivial for them – it reduces to the containment relation. The definition of the closure operator implies that A is closed in X if and only if $\bar{A} = A$. The relation between closed and open sets is very simple.

Proposition 3. *A set A is closed in a topological space $(X, \mathcal{T})$ if and only if $X \setminus A$ is open in $(X, \mathcal{T})$; that is, if and only if $X \setminus A \in \mathcal{T}$.*

Proposition 3 and the axioms of a topology $\mathcal{T}$ imply (by using de Morgan's laws) that the set of all closed subsets of a space X satisfy the following properties: the intersection of any collection of closed sets is closed, the union of a finite number of closed sets is closed, the empty set is closed and the whole space X is closed. It is easy to verify that the closure of a set A in a topological space X is equal to the intersection of all closed subsets of X which contain A. Consequently, $\bar{A}$ is the *smallest* closed set in X which contains A.

Example 7. Let $(X, \mathcal{T})$ be an antidiscrete space; that is, $\mathcal{T} = \{\varnothing, X\}$. Then the closure $\bar{A}$ of any nonempty subset $A \subset X$ is X, and the collection of all closed sets in X is exhausted by the sets $\varnothing$ and X. On the other hand, in a discrete space $(X, \mathcal{T} = \text{Exp } X)$, the nearness relation coincides with the containment relation, and the closure of any set is the same set. The closure of the set $\mathbb{Q}$ of all rational numbers in the Sorgenfrey line $(\mathbb{R}, \mathcal{T}_s)$ is $\mathbb{R}$. Nevertheless, recall that the space $(\mathbb{R}, \mathcal{T}_s)$ does not have a countable base.

The following elementary proposition is often useful.

Proposition 4. *If U and V are disjoint open subsets of X, then $\bar{U} \cap V = \varnothing$ (and $U \cap \bar{V} = \varnothing$).*

1.4. Definition of a Topology Using a Nearness Relation or a Closure Operator.

An *(abstract) closure operator* on X is a rule which assigns to each subset A of X a set $\bar{A} \subset X$ in such a way that conditions $1_c)$–$4_c)$ of the preceding section are satisfied.

If, for example, one is given an (abstract) nearness relation δ between points and subsets which satisfies conditions $1_\delta)$–$4_\delta)$, then setting

$$\bar{A} = \{x \in X : x \delta A\} \qquad \text{for each} \qquad A \subset X \tag{1}$$

defines a closure operator on X satisfying conditions $1_c)$–$4_c)$. An abstract closure operator determines a collection $\mathcal{F} = \{A \subset X : \bar{A} = A\}$ of sets which are closed with respect to the operator. It is not difficult to verify that the collection $\mathcal{T} = \{X \setminus A : A \in \mathcal{F}\}$ of complements of elements of $\mathcal{F}$ is a topology on X, that $\mathcal{F}$ is the collection of all closed subsets of the topological space $(X, \mathcal{T})$, that for any $A \subset X$ the closure $\bar{A}$ coincides with the closure of A in X, and the original relation δ is also the nearness relation generated by the topology $\mathcal{T}$: $\delta = \delta(\mathcal{T})$. This leads to the following important conclusion.

Proposition 5. *The topology of a space X is uniquely determined by any one of the following objects: the nearness relation δ, the closure operator, or the collection of all closed sets.*

1.5. Subspaces of a Topological Space.

To each subset Y of a topological space $(X, \mathcal{T})$ (that is, to each subset $Y \subset X$) is associated a new topological space $(Y, \mathcal{T}|_Y)$ where $\mathcal{T}|_Y = \{U \cap Y : U \in \mathcal{T}\}$ is the set of all "traces" in Y of the open subsets of X. This topology on Y is said to be the topology *generated* (or *induced*) by the topological space X.

The topological space $(Y, \mathcal{T}|_Y)$ is called a *subspace* of the space $(X, \mathcal{T})$. The simple device of passing to subspaces allows us to easily construct very different examples of topological spaces from the standard spaces that we have already seen: the line, the plane, n-dimensional euclidean space and so forth. It suffices to select a subset of one such and decree that it be considered as a subspace. The properties of a subspace can be very different from those of the whole space.

In view of the definition, any figure on the plane – a disk, a circle, a disk with holes, and so forth – is a topological space. The set $\mathbb{Q}$ of rational numbers and the set J of irrational numbers are topological spaces with topologies induced from the usual topology on the real line.

The *relative* nature of the concept of a closed set emerges in connection with the notion of a subspace. Indeed, if X is a topological space and Y a subspace, then a set $A \subset Y$ can be closed in Y but not closed in the whole space X. For example, the whole set Y is always closed in Y, but Y might not be closed in X. When a set A is considered simultaneously as a subset of different topological spaces, we denote its closure in a space Y by $\mathrm{Cl}_Y(A)$ instead of using an overbar. It is clear that $\mathrm{Cl}_Y A = (\mathrm{Cl}_X A) \cap Y$ when $A \subset Y \subset X$.

If the topology on Y is induced from X (so Y is a subspace of X), the nearness relation on Y coincides with the restriction to Y of the nearness relation on the original space X, and the closed subsets of Y are the "traces" in Y of the closed sets in X; that is, the closed subsets of Y are of the form $F \cap Y$ where F is closed in X.

Example 8 (The Cantor perfect set). This is one of the most interesting examples of spaces defined by passing to a subspace. First, some terminology: henceforth, we use the word *segment* (of the line $\mathbb{R}$) to refer to a set of the form $[a, b] = \{x \in \mathbb{R}: a \leqslant x \leqslant b\}$ and *interval* to refer to a set of the form $(a, b) = \{x \in \mathbb{R}: a < x < b\}$. If $A_n \subset \mathbb{R}$ is a union of a finite number of pairwise disjoint segments, we let A_{n+1} denote the set obtained by removing from each segment of A_n the middle interval equal to one third the length of the segment. We take A_1 to be the segment $[0, 1]$. This together with the construction of A_{n+1} from A_n gives a decreasing sequence $\{A_n: n \in \mathbb{N}^+\}$ of subsets of the line $\mathbb{R}$. The intersection $\bigcap_{n+1}^{\infty} A_n$ is a subset C of the line $\mathbb{R}$ called the *Cantor perfect set*. As a subspace of the line, the Cantor perfect set possesses a number of remarkable properties. We shall return to it again. For now, we point out only that C is closed in $\mathbb{R}$, that C has no isolated points, but does not contain any interval of $\mathbb{R}$, and that C is uncountable.

In closing, we record the following simple fact: if $\mathscr{B}$ is a base of a space X and if Y is a subspace of X, then $\{U \cap Y: U \in \mathscr{B}\}$ is a base of Y.

1.6. The Free Sum of Topological Spaces. Let $\{X_s: s \in S\}$ be a collection of topological spaces and let $\mathscr{T}_s$ denote the topology of X_s. Suppose first that the sets X_s are pairwise disjoint. Then $\mathscr{B} = \bigcup\{\mathscr{T}_s: s \in S\}$ is a base of a topology $\mathscr{T}$ on the set $X = \bigcup\{X_s: s \in S\}$ because $\mathscr{B}$ covers X and is closed with respect to finite intersection. The topological space $(X, \mathscr{T})$ is called the *free sum of the spaces* X_s over $s \in S$ and the topology $\mathscr{T}$ is called the *free sum of the topologies* $\mathscr{T}_s$ over $s \in S$. They are denoted by

$$X = \Sigma_{\oplus}\{X_s: s \in S\} \tag{2}$$

and

$$\mathscr{T} = \Sigma_{\oplus}\{\mathscr{T}_s: s \in S\}$$

respectively.

It is clear that each X_s is an open subspace of $X = \Sigma_{\oplus}\{X_s: s \in S\}$. Since the complement of X_s in X (that is, the set $\bigcup\{X_{s'}: s' \in S\setminus\{s\}\}$) is open, each X_s is closed in X. Thus, the subspaces X_s are disjoint open-closed "slices" of X.

If some of the sets X_s intersect, then the free sum can be defined up to canonical homeomorphism as follows. We first replace each space X_s by the canonically homeomorphic space $X_s' = X_s \times \{s\}$ with topology $\mathscr{T}_s' = \{U \times \{s\}: U \in \mathscr{T}_s\}$. The spaces X_s' are pairwise disjoint and we define the *free sum* $\Sigma_{\oplus}\{X_s: s \in S\}$ of the spaces X_s over $s \in S$ to be the topological space $\Sigma_{\oplus}\{X_s': s \in S\}$.

In this way, we can, for example, form the free sum of any collection $\{Y_s: s \in S\}$ of subspaces of a given space X. This idea plays an essential role in a number of constructions associated with maps (see [Ar2] and [ArP, chap. IV]).

Despite its simplicity, the operation of forming free sums is very useful for constructing examples: it is very convenient and especially important that any collection of topological spaces can be viewed as lying in a single topological space which preserves many topological properties of the members of the collection.

There is no operation having properties analogous to those of the free sum of topological spaces in the category of topological groups or the category of linear topological spaces or other topological algebraic categories.

1.7. Centered Collections of Sets and Convergence in Topological Spaces. The concept of a topological space allows one to give a very general mathematical expression to the fundamental idea of passage to a limit. The key notion, here, is that of convergence in a topological space, which evolved from the concept of a convergent sequence of real numbers.

Definition 6. A collection (system) ξ of subsets of a set X is said to be *centered* if the intersection of any finite number of its elements $A_1, \ldots, A_k$ is nonempty: $A_1 \cap \cdots \cap A_k \neq \varnothing$. It is clear that every element of a centered collection of sets must be nonempty.

If η is a collection of sets such that any two elements A, B are such that either $A \subset B$ or $B \subset A$, then η is said to be a *chain of sets*. It is clear that every chain of nonempty sets is a centered collection of sets.

The prototype of the concept of a centered collection of sets is a decreasing sequence of segments which arises in many basic arguments in mathematical analysis, notably in the proof of the Heine-Borel-Lebesgue lemma.

Let $(X, \mathscr{T})$ be a topological space and ξ a collection of subsets of X. A point $x \in X$ is said to be *adherent* to ξ if x belongs to the closure of each element of ξ: $x \in \bigcap \{\bar{A} \colon A \in \xi\}$. In this case we also say that the collection ξ is adherent to x. We now introduce the fundamental concept of convergence of a collection of sets to a point in a topological space. The definition relates only to centered collections of sets.

Definition 7. A centered collection ξ of subsets of a set X *converges* to a point $x \in X$ in a topological space $(X, \mathscr{T})$ if for each neighbourhood O_x of the point x there exists an element A of the collection ξ contained in O_x.

If $A_1 \in \xi$ and $\bar{A}_1 \not\ni x$, then $O_{1x} = X \setminus \bar{A}_1$ is a neighbourhood of the point x. If $A \in \xi$ satisfies the condition $A \subset O_x$, then $A \cap A_1 = \varnothing$ and the collection ξ is not centered. Consequently, if a centered collection ξ converges to a point x in a topological space X, then x is adherent to the collection ξ. The converse is not true.

Example 9. The collection ξ of all subsets of the line $\mathbb{R}$ whose complement is finite is centered. All points of $\mathbb{R}$ (in the usual topology) are adherent to ξ, but ξ does not converge to any point of $\mathbb{R}$.

Any nonempty centered collection of subsets of an antidiscrete space converges to every point of the space. By way of contrast, a centered collection ξ converges to a point $x \in X$ of a discrete space X if and only if $\{x\} \in \xi$.

The concept of a convergent centered system allows one to adequately express any phenomenon connected with convergence, continuity and passage to a limit. In contrast, the language of convergence of (countable) sequences is fundamentally limited (for example, it is impossible to define and describe the topology of a product of an uncountable set of topological spaces in terms of the language of countable sequences; see, also, the example in the following section). However, the notion of a convergent sequence can be defined in any topological space; in view of the exceptional simplicity and intuitive appeal of the definition, not to mention its classical nature, it is worth singling out the widest class of topological spaces in which topology and convergence can be described using convergent sequences.

Definition 8. A sequence $\{x_n\}_{n+1}^\infty$ of points of a topological space X is said to *converge to a point* $x \in X$ if for each neighbourhood O_x of the point x there exists a number $n \in \mathbb{N}^+$ such that $x_k \in O_x$ for all $k \geqslant n$; that is, if each neighbourhood of x contains all terms of the sequence after some fixed term.

1.8. Sequential Spaces. The Sequential Closure Operator

Definition 9. A topological space X is called *sequential* if, for every set $A \subset X$ which is not closed in X, there exists a sequence $\{x_n\}_{n=1}^\infty$ of points of A converging to a point of the set $\bar{A} \setminus A$.

Example 10. Let $X = \mathbb{R}^\mathbb{R}$ be the set of all real functions on the line $\mathbb{R}$. For $f \in X$, a positive number ε, and a finite set $K \subset \mathbb{R}$ we set $V(f, \varepsilon, K) = \{g \in X: |g(x) - f(x)| < \varepsilon$ for all $x \in K\}$. The collection $\mathscr{B}_p$ of all sets of the form $V(f, \varepsilon, K)$ is a base of a topology on X called the *topology of pointwise convergence* and denoted by $\mathscr{T}_p$. Let $A = C(\mathbb{R})$ be the set of all continuous real valued functions on $\mathbb{R}$. Then it is easy to check that $\bar{A} = X$.

We consider the set $B_1 = B_1(\mathbb{R})$ of all functions $g \in X$ for which there exists a sequence of elements of A converging to g. Thus B_1 is the set of functions of *first Baire class* on $\mathbb{R}$. It is known that $X \setminus B_1 \neq \varnothing$. Fix $g \in X \setminus B_1$ and consider the subspace $Y = A \cup \{g\}$ of the topological space X. The set A is not closed in Y since $g \notin A$. We have $\bar{A} \setminus A = \{g\}$ and no sequence of points in A converges to g. Hence the space Y is not sequential. The entire space X is not sequential either – the proof is not much more difficult.

Example 11. Consider the space $X = \mathbb{R}^\mathbb{R}$ in Example 10 and the set $S \subset X$ of all functions $g \in \mathbb{R}^\mathbb{R}$ which differ from being identically zero at no more than a countable set of points. Since a union of countably many countable sets is countable, the set S possesses the following property: the closure (in the space X) of any countable subset $M \subset S$ is contained in S. Thus, no sequence of points in S converges to a point of $X \setminus S$. But the set S is not closed because $\bar{S} = X \neq S$. Consequently, X is not a sequential space. In contrast, the subspace S of X can be shown to be sequential.

Although not every topological space is sequential, the sequential closure operator is defined on any topological space.

If A is any subset of a space X, let $[A]_{seq}$ be the set of all points x in X for which there exists a sequence in A converging to x. The set $[A]_{seq}$ is called the *sequential closure* of A in X and the map $A \to [A]_{seq}$ is called the *sequential closure operator* on the topological space X.

The sequential closure operator $[\ ]_{seq}$ on any space X possesses the following properties:

0_{seq}. $[\varnothing]_{seq} = \varnothing$;
1_{seq}. $A \subset [A]_{seq} \subset \bar{A}$ for all $A \subset X$;
2_{seq}. $[A \cup B]_{seq} = [A]_{seq} \cup [B]_{seq}$.

However, iteration of the sequential closure operator, generally speaking, gives a new set: the set $[[A]_{seq}]_{seq}$ can differ from the set $[A]_{seq}$.

Example 12. Let A be the subset of continuous functions in the space $X = \mathbb{R}^{\mathbb{R}}$ of all real-valued functions on $\mathbb{R}$ with the topology of pointwise convergence. Then, $[A]_{seq} = B_1$ is the set of all *functions of first Baire class* on $\mathbb{R}$ and $[[A]_{seq}]_{seq} = [B_1]_{seq} = B_2$ is the set of all *functions of second Baire class* on $\mathbb{R}$. Since $B_2 \neq B_1$, we have $[[A]_{seq}]_{seq} \neq [A]_{seq}$. Hence, the sequential closure operator is not idempotent; this contrasts to the closure operator on a topological space where one always has $\bar{\bar{A}} = \bar{A}$.

The space X in Example 12 is not sequential. It is important to note that it can also happen that $[[A]_{seq}]_{seq} \neq [A]_{seq}$ in a sequential space X.

Example 13. Let X be the set consisting of pairwise distinct objects of the following three types: points x_{mn} where $m \in \mathbb{N}^+$ and $n \in \mathbb{N}^+$, points y_n where $n \in \mathbb{N}^+$, and a point z. We set $V_k(y_n) = \{y_n\} \cup \{x_{mn}: m \geqslant k\}$ and let γ denote the set of subsets $W \subset X$ such that $z \in W$ and there exists a positive integer p such that $V_1(y_n) \setminus W$ is finite and $W \ni y_n$ for all $n \geqslant p$. The collection $\mathscr{B} = \{\{x_{mn}\}: n \in \mathbb{N}^+, m \in \mathbb{N}^+\} \cup \gamma \cup \{V_k(y_n): n, k \in \mathbb{N}^+\}$ is a base of a topology on X. In the space X, the sequence $\{x_{mn}\}_{m=1}^{\infty}$ converges to the point y_n and the sequence $\{y_n\}_{n=1}^{\infty}$ converges to the point z. Using the base $\mathscr{B}$ it is easy to check that X is a sequential space. However, for the set $A = \{x_{mn}: m \in N^+, n \in N^+\}$, we have $[A]_{seq} = A \cup B$ where $B = \{y_n: n \in N^+\}$ and $[B]_{seq} = B \cup \{z\}$. It follows that $[[A]_{seq}]_{seq} \ni z \notin [A]_{seq}$. Thus $[A]_{seq} \neq [[A]_{seq}]_{seq}$. It is clear that $[[A]_{seq}]_{seq} = X$.

Definition 10. A topological space X is called a *Frechet-Uryson space* if the closure of every subset $A \subset X$ in X coincides with the sequential closure of A: $\bar{A} = [A]_{seq}$.

It follows from Definitions 9 and 10 that every Frechet-Uryson space is a sequential space. Example 13 shows that not every sequential space is a Frechet-Uryson space.

Example 14. Consider the subspace $Y = A \cup \{z\}$ of the space X in example 13. It is easy to verify that no sequence of points in A converges to the point z. But $\bar{A} \setminus A = \{z\}$. Consequently, Y is not a sequential space. Thus a subspace of a sequential space need not be sequential.

Proposition 6. *A topological space X is a Frechet-Uryson space if and only if each subspace Y of X is sequential.*

1.9. The First Axiom of Countability and Bases of a Space at a Point (and at a Set)

Definition 11. A collection γ of open neighbourhoods of a point x in a topological space X is called a *base of the space X at the point x* if each neighbourhood of x contains an element of γ.

If the space X possesses a countable base at each point, then X is said to satisfy the *first axiom of countability*. All spaces with a countable base and all discrete spaces satisfy this axiom. We remark that a discrete space X has a countable base if and only if the set X is countable.

If a space X satisfies the first axiom of countability, then so does each of its subspaces. The following are easy to prove.

Proposition 7. *If X satisfies the first axiom of countability, then X is a Frechet-Uryson space.*

Proposition 8. *A countable space (that is, a space in which the set X is countable) satisfies the first axiom of countability if and only if it has a countable base.*

Example 15. The spaces X and Y in Examples 13 and 14 are countable, but do not satisfy the first axiom of countability: they are not even Frechet-Uryson spaces.

Thus, not every countable space has a countable base. Note that all single point sets (singletons) in Y are closed and only one point (the point z) is not isolated.

The condition that each point of a space X be the intersection of a countable collection of open subsets of X singles out a class of spaces which are called *spaces of countable pseudocharacter*. It is clear that the space Y of example 14 is such a space: $\{z\} = \bigcap \{X \setminus \{x_{mn}\}: n \in \mathbb{N}^+, m \in \mathbb{N}^+\}$. From this, we obtain the following.

Proposition 9. *Not every space of countable pseudocharacter is sequential; not every such space satisfies the first axiom of countability.*

Thus, it is possible to have a countable collection γ of open sets of a space X and a point $x \in X$ such that $\{x\} = \bigcap \gamma$ without X having a countable base at x.

Example 16. The Sorgenfrey line $(\mathbb{R}, \mathcal{T}_s)$ satisfies the first axiom of countability at all points: the countable collection $\{[a, a + 1/n)_{n=1}^{\infty}\}$ of open sets in $(\mathbb{R}, \mathcal{T}_s)$ is a base of $(\mathbb{R}, \mathcal{T}_s)$ at the point a.

The following generalization of the concept of a base of a space at a point is useful.

Definition 12. A family γ of open subsets of a topological space X is said to be a *base of X at the set $A \subset X$* if $A \subset \bigcap \gamma$ and, for each open set $U \subset X$ containing A, there exists $V \in \gamma$ such that $V \subset U$.

If $A \subset X$ is an infinite subset, then the space X does not, as a rule, have a countable base at A.

Example 17. Let X be the plane $\mathbb{R}^2$ with the usual topology and $A = \{(x, 0): x \in \mathbb{R}\}$. The space X does not possess a countable base at A.

In connection with Proposition 7, we remark that not every Frechet-Uryson space satisfies the first axiom of countability.

Example 18. Let A be an uncountable set and $\zeta \notin A$. Define a topology $\mathcal{T}_\zeta$ on the set $X = A \cup \{\zeta\}$ by setting $\mathcal{T}_\zeta = \{X \backslash K: K \subset X, K \text{ finite}\} \cup \{B: B \subset A\}$. Thus, all subsets of A are open in X and a set containing the point ζ is open if and only if its complement is finite. The space X is called the *Alexandrov supersequence* of the length $|A|$. It is clear that every sequence of pairwise distinct points of X converges to ζ. But X does not satisfy the first axiom of countability. In fact, if γ is a countable collection of open sets in X and $\bigcap \gamma \ni \zeta$, then the set $X \backslash \bigcap \gamma$ is countable (this follows from the way neighbourhoods of ζ were defined) and, hence, $(\bigcap \gamma) \cap A \neq \varnothing$. Fix $a \in A \cap (\bigcap \gamma)$. The neighbourhood $U = X \backslash \{a\}$ of ζ contains no element of γ.

The following example exhibits a countable Frechet-Uryson space with a single nonisolated point at which the first axiom of countability is not satisfied and, hence, at which the space does not have a countable base.

Example 19. Let $X = \{x_{mn}: n \in \mathbb{N}^+, m \in \mathbb{N}^+\}$ be a set of pairwise distinct points x_{mn} and adjoin a single point a which does not lie in X. Let $\mathcal{T}$ be the topology on $Y = X \cup \{a\}$ given by the base consisting of all subsets of X and all subsets V of Y for which the set $P_m = \{n \in \mathbb{N}^+: x_{mn} \notin V\}$ is finite for each $m \in \mathbb{N}^+$. All points of the space (Y, T) are isolated except the point a at which the first axiom of countability is not satisfied. It is clear that $(Y, \mathcal{T})$ is a countable Frechet-Uryson space. This space is called the *countable Frechet-Uryson fan*. Each sequence $\{x_{mn}, n \in \mathbb{N}^+\}$ converges in $(Y, \mathcal{T})$ to the point a.

1.10. Everywhere Dense Sets and Separable Spaces. Which subsets of a topological space should be considered "large" and which should be considered "small"? Since there is no concept of distance and volume in a topological space, the answer to this question is not completely clear. There are a number of possible answers: the most elementary involve the concepts of everywhere dense and nowhere dense sets.

Definition 13. A subset A of a topological space X is called *everywhere dense* in X if its closure is equal to X: $\bar{A} = X$.

In a sense, a set A which is everywhere dense in a space X approximates the space: every point $x \in X$ is near A, that is, of "distance zero" from A. In this sense an everywhere dense set is large.

It is clear that a subset $A \subset X$ is everywhere dense in X if and only if every nonempty open subset U of X contains at least one point of A (it suffices to require that this hold for all U in a base $\mathcal{B}$ of X).

A topological space is said to be *separable* if it contains a countable everywhere dense set.

Example 20. Every countable space is separable. The set $\mathbb{Q}$ of all rational numbers is dense in the space $\mathbb{R}$ of real numbers with the usual topology and in the Sorgenfrey line $(\mathbb{R}, \mathcal{T}_s)$. Hence these spaces are separable. The first has a countable base, but the second does not. Since $(\mathbb{R}, \mathcal{T}_s)$ satisfies the first axiom of countability, we obtain the following.

Proposition 10. *Not every separable space satisfying the first axiom of countability possesses a countable base.*

The following simple proposition is often used.

Proposition 11. *Every space with a countable base is separable.*

◁ Choose a point in each nonempty set in a countable base B of the space X. The set A of all such points is countable and everywhere dense in X. ▷

If a space has a countable base, so does every subspace of it. This is not the case with separability: a subspace of a separable space can fail to be separable (see [ArP. Ch. III]). Even if every subspace of a space X is separable, it does not follow that X has a countable base: the Sorgenfrey line (Example 6) or any countable space without a countable base (see Example 14) provides an example.

The following simple assertion is often useful.

Proposition 12. *If A is an everywhere dense subset of a topological space X and if U is open in X, then $\overline{A \cap U} = \overline{U}$.*

1.11. Nowhere Dense Sets. The Interior and Boundary of a Set. It is very temptating to consider a set to be small if it is the complement of a large set. Intuition certainly demands that this be the case. However, the complement of an everywhere dense set can be everywhere dense, that is, large. For example, the set $\mathbb{Q}$ of rational numbers and the set J of irrational numbers are both everywhere dense in the space of real numbers and are complements of one another in $\mathbb{R}$. We now define a notion which agrees much better with our intuition about when a set is "small".

Definition 14. A set $A \subset X$ is said to be *nowhere dense* in the topological space $(X, \mathcal{T})$ if, for each nonempty open set U, there exists a nonempty open set V such that $V \subset U$ and $V \cap A = \varnothing$. An equivalent condition is that the complement to the closure of A be everywhere dense in X: that is $\overline{X \setminus \overline{A}} = X$.

If a topological space X does not have isolated points and if each finite subset of X is closed, then any finite set $A \subset X$ is clearly nowhere dense in X. A straight line on the plane is a nowhere dense set. An important example of a nowhere dense set is the boundary of an open set.

Definition 15. The *boundary* $\mathrm{Bd}(A)$ *of a set* $A \subset X$ in a topological space X is the set of all points $x \in X$ which are near to both A and its complement $X \setminus A$. Thus, $\mathrm{Bd}(A) = (\overline{X \setminus A}) \cap \overline{A}$ and the set $\mathrm{Bd}(A)$ is always closed.

A set is closed if and only if it contains every point of its boundary. A set is open if and only if it has no point in common with its boundary. Consequently, a set is simultaneously open and closed (that is, *open-closed*) if and only if its boundary is empty. It is clear that we always have $\text{Bd}(A) = \text{Bd}(X \backslash A)$.

If both A and $X \backslash A$ are everywhere dense in the space X, then $\text{Bd}(A) = X = \text{Bd}(X \backslash A)$ is a very "large" set. This happens, for example, when $A = \mathbb{Q}$ and $X = \mathbb{R}$. The situation is different if A is an open set.

Proposition 13. *The boundary of any open set U in a topological space X is a nowhere dense set.*

◁ Suppose that $V \neq \varnothing$, V is open in X, and $V \subset \text{Bd}(U)$. Then $V \subset \bar{U}$ and $V \cap U \neq \varnothing$. But $(V \cap U) \cap (\overline{X \backslash U}) = \varnothing$ and since $V \cap U$ is open, we have $(V \cap U) \cap (X \backslash U) = \varnothing$. Hence, $(V \cap U) \cap \text{Bd}(U) = \varnothing$ contradicting the fact that $\varnothing \neq V \cap U \subset \text{Bd}(U)$. ▷

Definition 16. Let A be a subset of a topological space $(X, \mathscr{T})$. The *interior* of A in X is the set $\text{Int}(A)$ of all points of A for which A serves as a neighbourhood. Thus, $\text{Int}(A) = \{x \in A: \text{there exists } U \in \mathscr{T} \text{ such that } x \in U \subset A)$.

It can be shown that the interior of a set A is the union of all open sets contained in A.

The following simple relations hold: $\text{Int}(A) = A \backslash \text{Bd}(A)$ and $\text{Bd}(A) = \bar{A} \backslash \text{Int}(A)$. Nowhere dense sets can be characterized in terms of everywhere dense sets.

Proposition 14. *A set $A \subset X$ is nowhere dense in a topological space X if and only if the interior of its complement is everywhere dense in X; that is, $\overline{\text{Int}(X \backslash A)} = X$.*

In particular, a closed subset F of X is nowhere dense if and only if the interior of F is empty.

Proposition 14 shows that a subset of a topological space can reasonably be considered to be "large" if its interior is everywhere dense in X. A more refined approach to the definition of the concepts of large and small sets in a topological space will be given in §4.

1.12. Networks. There are some very useful generalizations of the concept of a base of a topological space. In this section we briefly sketch one of these – the concept of a network.

Definition 17. A collection S of subsets of a topological space X is called a *network of the space* (or a network in X) if, for each point $x \in X$ and each neighbourhood O_x of x, there exists $P \in S$ such that $x \in P \subset O_x$.

Thus, the only difference between a network and a base is that the elements of a network need not be open subsets.

It is clear that a collection S is a network in a topological space X if and only if every open subset of X is the union of some members of S.

Obviously every base of a topological space is a network. But the set $\{\{x\}; x \in X\}$ of all one point subsets of X is also a network of the space X. In particular, every countable space has a countable network, although not every countable set has a countable base (see Example 13).

Choosing a point from each nonempty element of a network gives an every-where dense subset of a space. Hence, we have the following.

Proposition 15. *A topological space with a countable network is separable.*

The converse is false: the Sorgenfrey line is a counterexample.

One of the main advantages of networks, as opposed to bases, is the following additivity property.

Proposition 16. *Let* $\{X_\alpha : \alpha \in A\}$ *be a collection of subspaces of a topological space X such that $X = \bigcup \{X_\alpha : \alpha \in A\}$. If S_α is a network of X_α for each $\alpha \in A$, then $S = \bigcup \{S_\alpha : \alpha \in A\}$ is a network of X.*

In particular, if S_α is a base of X_α for every α, then S is a notwork of X, but S may not be a base of X since the subsets X_α are not open in general.

§2. Some Important Classes of Topological Spaces

In this chapter, we consider topologies which arise naturally from different structures on sets such as metrics, orders, and so forth.

The last section of the chapter is devoted to separation axioms in topological spaces. In particular, we discuss which properties hold for the classes singled out below.

2.1. Linearly Ordered Spaces. Let $(X, <)$ be a linearly ordered set. We use the word "interval" to refer to sets of the form $(a, b) = \{x \in X : a < x < b\}$ where $a \in X, b \in X$ or $(\infty, a) = \{x \in X : x < a\}$ or $(a, \infty) = \{x \in X : a < x\}$ where $a \in X$.

The intersection of two intervals is an interval. Consequently, the collection of all intervals of a linearly ordered set $(X, <)$ is a base of a topology $\mathscr{T}$ on X: we say that this topology is generated by the linear ordering $<$.

Definition 1. A topological space $(X, \mathscr{T})$ is called a *linearly ordered space* if its topology is generated by a linear ordering on X.

The real line with the usual topology is an ordered space. However, the plane $\mathbb{R}^2$ with the usual topology is not an ordered space – this follows directly from connectedness considerations (see §8). Alternatively, one can base an argument on the following assertion.

Proposition 1. *A linearly ordered space possesses a base consisting of sets whose boundaries have no more than two points.*

◁ Consider the base consisting of all intervals. ▷

Linearly ordered spaces possess very special properties.

Proposition 2. *If a linearly ordered space X is sequential, then it satisfies the first axiom of countability.*

◁ Let $x \in X$, $x\delta(x, \infty)$ and $x\delta(\infty, x)$. There exist sequences $\{x_n\}_{n=1}^{\infty}$ and $\{y_n\}_{n=1}^{\infty}$ converging to x for which $x_n < x < y_n$ for all n. Then $\{(x_n, y_n)\}_{n=1}^{\infty}$ is a base of X at x. ▷

Warning. If A is an everywhere dense subset of a linearly ordered space X, it does not follow that the set of all intervals (a, b), (a, ∞), (∞, a) where $a, b \in A$ is a base of X. Moreover, the natural conjecture that arises from Proposition 2, namely that a separable, linearly ordered space necessarily has a countable base, is not true.

Example 1. Define a linear ordering of the Euclidean square $I^2 = [0, 1] \times [0, 1]$ by the rule: $(x_1, y_1) < (x_2, y_2)$ if and only $x_1 < x_2$ or $x_1 = x_2$ and $y_1 < y_2$. This is the so-called *lexicographic ordering on the square* I^2. Give I^2 the topology generated by this ordering and let X denote the resulting space. Consider the subspaces $Y_1 = \{(x, y) \in I^2: y = 0, x > 0\}$, $Y_2 = \{(x, y) \in I^2: y = 1, x < 1\}$ and $Z = Y_1 \cup Y_2$ of X. It can be shown that the topology generated by the lexicographic ordering on the set Z coincides with the topology of Z as a subspace of X. The space Z is separable (the countable set $A = \{(r, 0) \in I^2: r$ is rational, $r \neq 0\}$ is everywhere dense in Z), but Z does not have a countable base; in particular, the intervals (a, b) where $a, b \in A$ do not form a basis of Z. On Y_1, the lexicographic ordering coincides with the usual linear ordering: $(x_1, 0) < (x_2, 0) \Leftrightarrow x_1 < x_2$. Consequently, the topology on Y_1 generated by lexicographic ordering $<$ restricted to Y_1 is the same as its topology as a subspace of the usual real line. This topology is different from the topology on Y_1 as a subspace of the lexicographically ordered square X since each semi-interval of the form $(a, b] = c \in Y_1$: $a < c \leqslant b\}$ where $a, b \in Y_1$ is open in the space Y_1. As is easy to check, the spaces Y_1 and Y_2 can be identified with semi-intervals of the Sorgenfrey line (see example 6 of §1).

Proposition 3. *The topology of the Sorgenfrey line is not generated by any linear ordering on $\mathbb{R}$.*

Example 1 and Proposition 3 easily imply the following.

Proposition 4. *A subspace of a linearly ordered space need not be a linearly ordered space.*

Example 2. *A discrete space X is linearly ordered.*

◁ Suppose X is infinite. Choose pairwise disjoint countably infinite subsets X_α of X such that: $X = \bigcup\{X_\alpha: \alpha \in A\}$. Order each X_α linearly as the set of integers. We also fix a linear ordering on A. If $\alpha' < \alpha''$, we will deem that any point $x' \in X_{\alpha'}$ precedes any point $x'' \in X_{\alpha''}$. Every point x of this linearly ordered space X has a nearest point to the "left" and a nearest point to the "right". Consequently the topology generated by the linear ordering is discrete. ▷

2.2. Metric Spaces. A *metric* ρ on a set X is a map which associates to each pair of points in X a non-negative real number $\rho(x, y)$ such that the following (metric) axioms are satisfied:

1_ρ. $\rho(x, y) = 0 \Leftrightarrow x = y$, for all $x, y \in X$;

2_ρ. $\rho(x, y) = \rho(y, x)$ for all $x, y \in X$;

3_ρ. $\rho(x, z) \leqslant \rho(x, y) + \rho(y, z)$ for all $x, y, z \in X$.

The pair (X, ρ) is then called a *metric space* and the number $\rho(x, y)$ is called the *distance between the points* x and y. The metric axioms express in abstract form the well known properties of the usual distance in the plane and in three dimensional space. Condition 3 is called the *triangle axiom* and condition 2 the *symmetry axiom*.

Each metric space (X, ρ) has a unique topology $\mathscr{T}_\rho$ generated by the metric ρ, a base of which is the collection $\mathscr{B}_\rho$ of spherical neighbourhoods of points of X: $\mathscr{B}_\rho = \{O_a(x) : x \in X, a \in \mathbb{R}, a > 0\}$ where $O_a(x) = \{y \in X : \rho(x, y) < a\}$. It is not difficult to verify that $\mathscr{B}_\rho$ is in fact a base – see §1. The notation $O_a(x)$ will be retained in what follows.

The *distance in a metric space* (X, ρ) *from a point* $x \in X$ *to a set* $A \subset X$ is defined by the formula $\rho(x, A) = \inf\{\rho(x, y) : y \in A\}$. The *distance* $\rho(A, B)$ *between two subsets A and B of X* is defined to be $\rho(A, B) = \inf\{\rho(x, y) : x \in A \text{ and } y \in B\}$.

The closure operator in a topological space $(X, \mathscr{T}_\rho)$ is easy to describe directly: for each $A \subset X$, $\overline{A} = \{x \in X : \rho(x, A) = 0\}$. A set A is closed in $(X, \mathscr{T}_\rho)$ if and only if each point which does not belong to A is a positive distance from A.

Example 3. Let X be an arbitrary set and define a metric ρ by the rule: $\rho(x, y) = 1$ if and only if $x \neq y$. This metric is said to be *trivial* and generates the discrete topology on X.

Example 4. Define the metric ρ on the set $\mathbb{R}$ of real numbers by the rule: $\rho(x, y) = |x - y|$ for all $x, y \in \mathbb{R}$. This is called the *usual metric* on $\mathbb{R}$; it generates the usual topology $\mathscr{T}_\rho$ on $\mathbb{R}$ which coincides with the topology generated by the linear ordering of the real numbers by magnitude (see Section 2.1).

Example 5. The *euclidean metric* ρ on the set $\mathbb{R}^n$ of ordered n-tuples $\bar{x} = (x_1, \ldots, x_n)$ of real numbers is defined by the formula: $\rho(\bar{x}, \bar{y}) = \sqrt{\sum_{i=1}^n (x_i - y_i)^2}$ where $\bar{x} = (x_1, \ldots, x_n) \in \mathbb{R}^n$ and $\bar{y} = (y_1, \ldots, y_n) \in \mathbb{R}^n$. The topology generated by this metric is called the *usual* topology on $\mathbb{R}^n$.

Example 6. Let H be the set of all infinite sequences $\{x_i\}_{i=1}^\infty$ such that $\sum_{i=1}^n x_i^2 < \infty$. For any $x = \{x_i\}_{i=1}^\infty$ and $y = \{y_i\}_{i=1}^\infty$, we set $\rho(x, y) = \sqrt{\sum_{i=1}^\infty (x_i - y_i)^2}$. The metric space (H, ρ) is called a *Hilbert space* (of countable weight). The topological space $(H, \mathscr{T}_\rho)$ has a countable base: it plays the role of a *universal space* for spaces with a countable base (see Section 2.7).

Example 7. Let A be a set, τ its cardinality, and $H(A)$ the set of all nonnegative real functions z on A such that $z(\alpha) \neq 0$ for at most countable many points $\alpha \in A$ and the series $\sum_{\alpha \in A} (z(\alpha))^2$ converges. Following tradition, we write z_α in place of $z(\alpha)$ for $z \in H(A)$.

The set $H(A)$ together with the metric ρ defined for $z, z' \in H(A)$ by the rule $\rho(z, z') = \sqrt{\sum_{\alpha \in A}(z_\alpha - z'_\alpha)^2}$ is called a *generalized Hilbert space of weight τ* and denoted $H(\tau)$ (the construction does not depend in an essential way on the choice of a set A of given cardinality τ).

Example 8. Let X be any nonempty set. We introduce a metric ρ on the set $\mathscr{F}(X)$ of all bounded real functions on X by the formula

$$\rho(f, g) = \sup_{x \in X} |f(x) - g(x)|$$

Convergence of sequences in $(\mathscr{F}(X), \mathscr{T}_\rho)$ is called *uniform convergence*. The space $(\mathscr{F}(X), \mathscr{T}_\rho)$ and its different subspaces, considered as metric spaces or as topological spaces, occur in many applications in functional analysis and general topology (see [Ar6], [Ru]).

Example 9. Let S be an infinite set of cardinality τ and, for each $s \in S$, set $I_s = I \times \{s\}$ for $s \in S$ where $I = (0, 1]$ is the unit semi-interval. To the set $P = \bigcup\{I_s : s \in S\}$, we adjoin a new point $(0, 0)$. We set $\tilde{P} = \{(0, 0)\} \cup P$ and define a metric ρ on $\tilde{P}$ as follows: if $(x, t_1), (y, t_2) \in \tilde{P}$, then $\rho((x, t_1), (y, t_2)) = |x - y|$ when $t_1 = t_2$ and $\rho((x, t_1), (y, t_2)) = |x| + |y|$ when $t_1 \neq t_2$. We denote the metric space $(\tilde{P}, \rho)$ by $J(\tau)$ and call it the *hedgehog space of spininess τ*.

The hedgehog $J(\tau)$ plays an essential role in a number of general arguments and constructions. In particular, a metric space of given weight τ can be represented as a subspace of the product of countably many copies of $J(\tau)$ of spininess τ (*Kowalsky's theorem*, see [Ko]).

Example 10. Let τ be any cardinal and $D(\tau)$ a set of cardinality τ. We let $B(\tau)$ denote the set of all sequences $\bar{x} = \{x_i\}_{i=1}^{\infty}$ with $x_i \in D(\tau)$ for $i \in \mathbb{N}^+$. For any $\bar{x} = \{x_i\}_{i=1}^{\infty}$ and $\bar{y} = \{y_i\}_{i=1}^{\infty}$ in $B(\tau)$, we set $\rho(\bar{x}, \bar{y}) = 1/k$ where k is the least integer i such that $x_i \neq y_i$. This determines a metric ρ on the set $B(\tau)$. The metric space $(B(\tau), \rho)$ is called the *Baire space of weight τ*. The topological space $(B(\tau), \mathscr{T}_\rho)$ has a base consisting of open-closed sets.

Remark. Let (X, ρ) be a metric space. Each subset $Y \subset X$ can itself be considered as a metric space by restricting the metric ρ to Y. The topology of the metric space (Y, ρ) coincides with the subspace topology on Y generated by $(X, \mathscr{T}_\rho)$.

Definition 2. Two metrics ρ_1 and ρ_2 on a set X are called *equivalent* if they generate the same topology on X; that is, if $\mathscr{T}_{\rho_1} = \mathscr{T}_{\rho_2}$.

A metric ρ on a set X is said to be *bounded* if there exists a real number $a > 0$ such that $\rho(x, y) < a$ for all $x, y \in X$. If ρ is a bounded metric on X, then we can define the *diameter* $\mathrm{diam}(A)$ of a set $A \subset X$ to be $\mathrm{diam}(A) = \sup(\rho(x, y) : x, y \in A\}$.

Proposition 5. *Any metric ρ on a set X is equivalent to a bounded metric ρ_1 on X in which the diameter of X does not exceed 1.*

◁ The function ρ_1 defined by the rule $\rho_1(x, y) = \min(1, \rho(x, y))$ for all $x, y \in X$ is the desired metric. ▷

Let ρ_1 and ρ_2 be two metrics on a set X. We write $\rho_1 \leqslant \rho_2$ and say that ρ_2 is *stronger than* ρ_1 if $\rho_1(x, y) \leqslant \rho_2(x, y)$ for all $x, y \in X$.

Proposition 6. *If a metric ρ_2 is stronger than a metric ρ_1, then the topology $\mathcal{T}_{\rho_2}$ generated by ρ_2 is stronger than the topology $\mathcal{T}_{\rho_1}$ generated by ρ_1; that is, $\mathcal{T}_{\rho_1} \subset \mathcal{T}_{\rho_2}$.*

2.3. Metrizable Spaces

Definition 3. A topological space $(X, \mathcal{T})$ is called *metrizable* if there exists a metric ρ on S which generates the topology $\mathcal{T}$; that is, if $\mathcal{T}_\rho = \mathcal{T}$.

It is clear that any subspace of a metrizable space is metrizable.

Proposition 7. *Each metrizable space satisfies the first axiom of countability.*

◁ The collection $\{O_{1/n}(x)\}_{n=1}^{\infty}$ of sets $O_{1/n}(x) = \{y \in X: \rho(x, y) < 1/n\}$ is a base of the space $(X, \mathcal{T}_\rho)$ at the point x. ▷

Therefore, the closure operator in a metrizable space $(X, \mathcal{T}_\rho)$ can be described in terms of sequences: $x \in \bar{A}$ if and only if there exists a sequence $\{x_n\}_{n=1}^{\infty}$ of points in A converging to x. Moreover, the convergence of $\{x_n\}_{n=1}^{\infty}$ to x can be described in terms of the metric ρ generating the topology $\mathcal{T}_\rho$ on X as follows: $\lim_{n \to \infty} \rho(x_n, x) = 0$.

Any one point subset and, hence, any finite subset, of a metrizable space is closed. In particular, the antidiscrete space is not metrizable. The following assertion allows us to exhibit examples of spaces which satisfy the first axiom of countability and many additional restrictions, but which are nevertheless not metrizable.

Proposition 8. *Every separable metric space X possesses a countable base.*

◁ Let ρ be a metric on X which generates the topology on X. Fix a countable dense subset A of X. The collection $\mathcal{B} = \{O_{1/n}(x): x \in A, n \in \mathbb{N}^+\}$ is countable and consists of open subsets of X. We shall show that $\mathcal{B}$ is a base of the space X. Let $x \in X$ and U be an open subset of X such that $x \in U$. Then $a = \rho(x, X \setminus U) > 0$. Choose $n \in \mathbb{N}^+$ so that $1/n < a/3$. Fix a point y of A in the open set $O_{1/n}(x)$. Then $x \in O_{1/n}(y) \in \mathcal{B}$ and, by the triangle axiom, $O_{1/n}(y) \subset U$. ▷

The Sorgenfrey line $(\mathbb{R}, \mathcal{T}_s)$ (see §1, Example 6) is a nonmetrizable separable space satisfying the first axiom of countability. That it is not metrizable follows from Proposition 8 because $(\mathbb{R}, \mathcal{T}_s)$ does not have a countable base.

The general sufficient conditions for a space to be metrizable are nontrivial: they are based on ideas connected with paracompactness and Stone's deep theorem about the paracompactness of every metric space (see [St1], [ArP], [E]) It is simpler to formulate the condition for metrizability when the space has a countable base (see [U]).

2.4. Prametrics, Symmetrics, and the Topologies they Generate.

The impetus to describe a topology using a distance function is justified not only by the intuitive nature of distance considerations, but also because it allows one to use the apparatus of the real numbers and, in particular, their continuity and order properties in investigations. In order to encompass the widest range of topolgical spaces, one can either omit one of the metric axioms or consider metrics with values in structures different from the field $\mathbb{R}$ of real numbers. In this section, we take some steps in the first direction.

Definition 4. A *prametric* on a set X is a map ρ of the set $X \times X$ to the set $\mathbb{R}^+$ satisfying the condition $\rho(x, x) = 0$ for all $x \in X$. A set X together with a prametric ρ on X is called a *prametric space* (X, ρ). The number $\rho(x, y)$ is called the *distance between the points* x and y in (X, ρ); we do not exclude the possibility that $x \neq y$ and $\rho(x, y) = 0$, or that $\rho(x, y) \neq \rho(y, x)$. The *distance* $\rho(x, A)$ *from a point* $x \in X$ *to a set* $A \subset X$ is $\inf\{\rho(x, y): y \in A\}$.

An arbitrary prametric ρ on a set X generates a topology $\mathcal{T}_\rho$ on X as follows: a set $F \subset X$ is closed if and only if the distance $\rho(x, F)$ is greater than zero for all $x \in X \setminus F$. Equivalently, a subset $U \subset X$ is open (that is, $U \in \mathcal{T}_\rho$) if and only if for each point $x \in U$ there exists a number $\varepsilon > 0$ such that $O_\varepsilon(x) \subset U$ where $O_\varepsilon(x) = \{y \in X: \rho(x, y) < \varepsilon\}$ is the ε-ball with center at x.

Warning. The collection $\{O_\varepsilon(x): x \in X, \varepsilon > 0\}$ might not be a base of the topology $\mathcal{T}_\rho$: not only can a set $O_\varepsilon(x)$ fail to be open, but its interior Int $O_\varepsilon(x)$ might not contain the point x. In fact, the set Int $O_\varepsilon(x)$ can even be empty. The topologies generated by prametrics can differ radically from the topologies generated by metrics.

A prametric is said to be *separating* if $\rho(x, y) = 0$ implies $x = y$ for $x, y \in X$.

A prametric on a set X is called *symmetric* if $\rho(x, y) = \rho(y, x)$ for all $x, y \in X$. A symmetric separating prametric is called a *symmetric*.

Example 11. Let (X, ρ) be a metric space and $\{P_\alpha: \alpha \in A\}$ be a collection of subsets of X. Set $d(\alpha_1, \alpha_2) = \rho(P_{\alpha_1}, P_{\alpha_2})$ for any $\alpha_1, \alpha_2 \in A$. It is clear that d is a symmetric prametric on the set X. If all the sets P_α are finite and pairwise disjoint, then d is a symmetric on the set A which is not, generally speaking, a metric.

Symmetric prametrics arise very naturally in the situation sketched in Example 12 involving mappings of metric spaces.

Example 12. Let (X, ρ) be a metric space and $f: X \to Y$ a map onto a set Y. Set $d(y_1, y_2) = (f^{-1}(y_1), f^{-1}(y_2))$ for all $y_1, y_2 \in Y$. Then d is a symmetric prametric on the set Y. Of particular importance is the case when d turns out to be a symmetric on Y.

The question of how the properties of the topology $\mathcal{T}_d$ on Y generated by the prametric d are related to the properties of the topology $\mathcal{T}_\rho$ generated by the metric ρ is nontrivial.

Example 13. Define a prametric d on the set $D = \{0, 1\}$ by the conditions: $d(0, 1) = 1$ and $d(1, 0) = 0$. The prametric d generates the *connected two point set* topology $\mathcal{T}_d = \{\varnothing, \{0\}, \{0, 1\}\}$ on D.

It is clear that a symmetric on a finite set generates the discrete topology.

Example 14. Let $X = \{x_{mn}: n \in \mathbb{N}^+, m \in \mathbb{N}^+\}$ and $Y = \{y_n: n \in \mathbb{N}^+\}$ where all x_{mn} are distinct and all y_n are distinct, $X \cap Y = \varnothing$, and $0 \notin X \cup Y$. Define a symmetric ρ on the set $Z = \{0\} \cup X \cup Y$ by the conditions $\rho(y_n, x_{mn}) = 1/m$ for all n, $m \in \mathbb{N}^+$, $\rho(0, y_n) = 1/n$ for $n \in \mathbb{N}^+$, and $\rho(z_1, z_2) = 1$ for all other pairs of distinct points of the set Z (with due regard for the symmetry axiom). The topology $\mathcal{T}_\rho$ generated by the symmetric ρ makes the set Z into a topological space identical to the space of example 13 of §1. For $\varepsilon = 1/k$, $k \in \mathbb{N}^+$, we have $O_\varepsilon(0) = \{0\} \cup \{y_n: n > k\}$. The interior Int $O_\varepsilon(0)$ does not contain the point 0 since $0 \in \bar{X}$ but $O_\varepsilon(0) \cap X = \varnothing$. We remark that $\rho(0, X) = 1$ although $0 \in \bar{X}$. This gives rise to the following cautionary remark.

Warning. Let X be a set on which the topology is generated by a symmetric ρ and let $x \in X$ and $x \in \bar{A}$. It does not follow that $\rho(x, A) = 0$. On the other hand, if $\rho(x, A) = 0$ it easily follows that $x \in \bar{A}$.

The space in Example 14 does not satisfy the first axiom of countability (at the point 0). Thus, the topology generated by a symmetric need not satisfy the first axiom of countability. In connection with this, it is useful to note the following.

Proposition 9. *If the topology of a space X is generated by a prametric ρ, then the space is sequential.*

◁ Suppose that A is a set which is not closed in X. Then there exists a point $x \in X \backslash A$ for which $\rho(x, A) = 0$. The sequence $\{a_n\}_{n=1}^\infty$ where $a_n \in A$ and $\rho(x, a_n) < 1/n$ converges to the point x of X. ▷

Thus, if X is not a sequential space, its topology cannot be generated by a symmetric on X.

Not only does a prametric ρ on X define a topology and, hence, a closure operator on subsets of X, but it also defines a new, more refined, restricted closure operator, the *praclosure operator*, on the sets $A \subset X$ in (X, ρ). Namely the *praclosure of a set A in a prametric space* (X, ρ) is defined to be the set $[A]_\rho = \{x \in X: \rho(x, A) = 0\}$.

For the space (Z, ρ) in Example 14 we have: $[X]_\rho = X \cup Y \neq Z = [[X]_\rho]_\rho$. Thus, in contrast to the closure operator, the praclosure operator does not, generally speaking, satisfy the condition $[[A]_\rho]_\rho = [A]_\rho$.

For any prametric ρ on X the praclosure operator satisfies the conditions:
1°. $[\varnothing]_\rho = \varnothing$;
2°. $A \subset [A]_\rho \subset \bar{A}$ for all $A \subset X$;
3°. $[A \cup B]_\rho = [A]_\rho \cup [B]_\rho$ for all $A, B \subset X$.

Proposition 10. *If $[A]_\rho = A$. then the set A is closed in $(X, \mathcal{T}_\rho)$. If $[[A]_\rho]_\rho = [A]_\rho$, then $[A]_\rho = \bar{A}$ (and conversely). If $[[A]_\rho]_\rho = [A]_\rho$ for all $A \subset X$, then the collection $\{\text{Int } O_{1/n}(x)\}_{n=1}^\infty$ is a base of $(X, \mathcal{T}_\rho)$ at the point $x \in X$ for all $x \in X$ and, consequently, the space $(X, \mathcal{T}_\rho)$ satisfies the first axiom of countability.*

A prametric ρ on X is called *strong* if $[A]_\rho = \bar{A}$ (that is, $[[A]_\rho]_\rho = [A]_\rho$) for all $A \subset X$ (where $\bar{A}$ is the closure of the set A in the topological space $(X, \mathcal{T}_\rho)$).

Prametrics which are not symmetric arise naturally in connection with linear orderings.

Example 15. The following defines a separating prametric ρ^* on the real line $\mathbb{R}$:

$$\rho^*(x, y) = |x - y| \text{ if } x \leqslant y,$$

$$\rho^*(x, y) = 1 \text{ if } y < x.$$

It is easy to check that the set $\mathbb{R}$ endowed with the topology $\mathcal{T}_{\rho^*}$ generated by ρ^* is the Sorgenfrey line $(\mathbb{R}, \mathcal{T}_s)$ (see § 1, Example 6). It is curious that there does not exist a symmetric on $\mathbb{R}$ which generates the topology of the Sorgenfrey line.

In the last example of this section, we mention another typical situation in which symmetrics arise naturally from metrics.

Example 16. Let ρ_1 and ρ_2 be two metrics on a set Z. For every $z_1, z_2 \in Z$ we set

$$d(z_1, z_2) = \min\{\rho_1(z_1, z_2), \rho_2(z_1, z_2)\}.$$

It is clear that d is a symmetric and it is easy to cite examples where d is not a metric (because the triangle axiom fails). It is also clear that a subset $A \subset Z$ is closed in the space $(Z, \mathcal{T}_d)$ if and only if it is closed in both $(Z, \mathcal{T}_{\rho_1})$ and $(Z, \mathcal{T}_{\rho_2})$. Thus, the topology $\mathcal{T}_d$ on Z generated by the symmetric d is the intersection of the metric topologies $\mathcal{T}_{\rho_1}$ and $\mathcal{T}_{\rho_2}$.

Now consider the set $Z = \{0\} \cup X \cup Y$ of Example 14. Suppose, in addition, that $x_{mn} = (1/n, 1/m)$ and $y_n = (1/n, 0)$ are points of the plane $\mathbb{R}^2$. Define metrics ρ_1 and ρ_2 on Z as follows. If $z_1, z_2 \in X$ or $z_1, z_2 \in Y$, take $\rho_1(z_1, z_2)$ and $\rho_2(z_1, z_2)$ to be the usual distance between the points z_1 and z_2 in the Euclidean metric on the plane $\mathbb{R}^2$ (see Section 2.3). If $z_1 \in X$ and $z_2 \in Y$, take $\rho_1(z_1, z_2)$ to be again equal to the distance between z_1 and z_2 on the Euclidean plane $\mathbb{R}^2$ and $\rho_2(z_1, z_2) = 1$. Set $\rho_1(0, z) = 1$ for all $z \in X \cup Y$, $\rho_2(0, z) = 1$ for all $z \in X$ and $\rho_2(0, y_n) = 1/n$. The symmetric d on Z defined by the rule: $d(z_1, z_2) = \min\{\rho_1(z_1, z_2), \rho_2(z_1, z_2)\}$ for $z_1, z_2 \in Z$ generates the topology on the set Z which was defined in Example 14 (see also § 1, Example 13).

Thus a topology which is the intersection of two metrizable topologies need not be a metrizable topology.

2.5. Abstract Praclosure Operators

Definition 5. An (abstract) *praclosure operator* $[\ ]_p$ on a set X is a map which associates to each set $A \subset X$ a set $[A]_p$ such that:

1_p) $[\varnothing]_p = \varnothing$;

2_p) $A \subset [A]_p$ for each $A \subset X$;

3_p) $[A \cup B]_p = [A]_p \cup [B]_p$ for all $A \subset X$ and $B \subset X$.

We shall often write $[A]$ in place of $[A]_p$.

One obtains a praclosure operator, for example, from a nearness relation δ between points and subsets of X which satisfies the following constraints for all $x \in X$, $A \subset X$, $B \subset X$: 1_δ) $x\bar{\delta}\varnothing$; 2_δ) $x \in A \Rightarrow x\delta A$; and 3_δ) if $x\delta(A \cup B)$, then $x\delta A$ or $x\delta B$. We set $[A]_p = \{x \in X : x\delta A\}$.

In this section, we sketch a method for topologizing a set which arises in many concrete situations and which uses the notion of a praclosure operator. The method provides a new approach to analyzing closure operators on topological spaces (see [C2], [G], [ArF]).

Definition 6. Let $[\ ]_p$ be a praclosure operator on a set X. A set $A \subset X$ is said to be *closed* if $[A]_p = A$. A set $U \subset X$ is said to be *open* if the set $A = X \setminus U$ is closed.

It is easy to check that the collection of all open sets is a topology $\mathcal{T}_p$ on X which we shall say is *generated by the praclosure operator* $[\ ]_p$.

The closure operator on a topological space $(X, \mathcal{T}_p)$ satisfies the condition: $[A]_p \subset \bar{A}$ for all $A \subset X$.

Example 17. Let $(X, \mathcal{T})$ be a topological space. If $A \subset X$, let $[A]_{\text{seq}}$ be the set of all points to which some sequence of points of the set A converges. Then $[\ ]_{\text{seq}}$ is a praclosure operator on X and the space $(X, \mathcal{T})$ is sequential if and only if the topology $\mathcal{T}$ is generated by the operator $[\ ]_{\text{seq}}$ in the sense of Definition 6; that is, if $\mathcal{T} = \mathcal{T}_{\text{seq}}$.

Example 18. Let ρ be a prametric on a set X and $[A]_p = \{y \in X : \rho(y, A) = 0\}$ for each $A \subset X$. Then $[\ ]_p$ is a praclosure operator on X and the topology generated by $[\ ]_p$ coincides with the topology generated by ρ on X.

One approach to the investigation of the closure operator on a topological space generated by a praclosure operator is based on the following.

Let $\tau = |X|$ be the cardinality of the set X, $\tau \geqslant \aleph_0$, and let τ^+ be the least cardinal greater than τ. Let $[\ ]_p$ be a praclosure operator on X. For all $A \subset X$ and all ordinals $\alpha \leqslant \tau^+$ (see [12, Chapter 1]), we define the set $[A]_p^\alpha$ by transfinite recursion as follows.

We set $[A]_p^0 = A$ and $[A]_p^\alpha = \bigcup_{\beta < \alpha} [A]_p^\beta$ if α is a limit ordinal. If $\alpha = \beta + 1$, then we set $[A]_p^\alpha = [[A]_p^\beta]_p$.

Proposition 11. *If the topology of X is generated by a praclosure operator $[\ ]_p$, then for each $A \subset X$ the closure $\bar{A}$ in the space X can be expressed as $\bar{A} = \bigcup_{\alpha < \tau^+} [A]_p^\alpha = [A]_p^{\tau^+}$ where τ is the cardinality of X.*

The following sharpening of Proposition 11 pertaining to sequential spaces is very useful.

Proposition 12. *If a topological space $(X, \mathcal{T})$ is sequential, then for each $A \subset X$*

$$\bar{A} = \bigcup_{\alpha < \aleph_1} [A]_{\text{seq}}^\alpha = [A]_{\text{seq}}^{\aleph_1}.$$

Here, the cardinality of X does not play a role.

An identical formula also holds for a praclosure operator generated by a prametric.

Warning. In a countable sequential space $(X, \mathcal{T})$ the set $[A]_{\text{seq}}^{\aleph_0}$ may be different from the closure of A. Moreover, there might not exist $\alpha^* < \aleph_1$ for which $\bar{A} = [A]_{\text{seq}}^{\alpha^*}$ for all $A \subset X$. This is proved in the paper [ArF] of Arkhangel'skiĭ and Franklin.

2.6. The Separation Axioms T_0, T_1 and T_2. What kinds of conditions are needed to ensure that disjoint subsets A and B of X be separated by disjoint open subsets; that is, when do there exist open subsets U and V of X such that $A \subset U$, $B \subset V$ and $U \cap V = \emptyset$? It is clear that this is not always the case: in the antidiscrete topology the only nonempty open set is the entire set X. In a metric space, it follows from the triangle axiom that any two disjoint closed sets can be separated by disjoint open sets.

The separation axioms postulate the possibility of separating sets A and B under different sorts of natural restrictions on A and B. Intuitively, these axioms should be thought of as abstractly expressed corollaries of the metric axioms – especially the triangle axiom.

The first three basic separation axioms T_0, T_1, and T_2 of a topological space pertain to the separation of points from one another; that is, they relate to the case when A and B are one point sets.

Axiom T_0. For any two distinct points x, y of the topological space X, there exists an open set U containing exactly one of these points; that is, such that $|U \cap \{x, y\}| = 1$.

Axiom T_1. For any two distinct points x, y of the topological space X, there exists an open set U such that $x \in U$ and $y \notin U$.

Axiom T_2. Any two distinct points x, y of the space X can be separated by disjoint neighbourhoods; that is, there exist open subsets U and V of X such that $x \in U$, $y \in V$ and $U \cap V = \emptyset$.

It is clear that $T_2 \Rightarrow T_1 \Rightarrow T_0$. An antidiscrete space X containing more than one point does not satisfy the axiom T_0.

Example 19. Let $X = \{0, 1\}$ be the two point set and $\mathcal{T} = \{\emptyset, \{0\}, \{0, 1\}\}$. The space $(X, \mathcal{T})$ (the connected two point space) satisfies the axiom T_0, but does not satisfy T_1: there does not exist an open set U containing the point 1 and not containing the point 0. This example is a special case of the following example which arises naturally in combinatorial topology.

Example 20. Let S be a simplex. We define a closure operator on the set X of all faces of S (including S). Let closure of a point t in X be the set of all faces of the simplex t and define the closure of an arbitrary subset $A \subset X$ to be the union of the closures of all points in A. This closure operator defines a topology on X which satisfies axiom T_0 but not T_1.

The following example is important for algebraic geometry (see [B4]).

Example 21. Let A be a commutative ring with a unit and X the set of all prime ideals of A. For any $a \in A$, let X_a denote the set of all prime ideals in A which do not contain a. It is clear that $X_a \cap X_b = X_{ab}$ for all $a, b \in X$, $X_0 = \emptyset$ and $X_1 = X$. Consequently, the collection $\mathcal{B} = \{X_a : a \in A\}$ is a base of a topology $\mathcal{T}$ on A. This topology is called the *spectral* or *Zariski topology*.

The topological space $(X, \mathcal{T})$ is called the *prime spectrum of the ring A* and is denoted by Spec(A). One can show that the closure of a one point set $\{x\}$ in Spec(A) consists of all prime ideals $y \in X = \text{Spec}(A)$ containing x. It follows that the space $(X, \mathcal{T})$ satisfies the separation axiom T_0, but not T_1, since the only closed points in X are the maximal ideals of the ring A.

Example 22. Let X be an infinite set and let the topology $\mathcal{T}$ consist of the empty set and all subsets of X whose complements are finite. Any two nonempty open sets in this space intersect. At the same time all one point sets in $(X, \mathcal{T})$ are closed. Hence the space $(X, \mathcal{T})$ satisfies axiom T_1 but not axiom T_2.

Proposition 13. *A topological space X satisfies the separation axiom T_1 if and only if all finite subsets of X are closed in X.*

In particular, a finite topological space satisfies axiom T_1 if and only if it is discrete. Example 19 shows that this is not the case for axiom T_0.

A space which satisfies axiom T_2 is also called a *Hausdorff space* and the axiom T_2 itself is called the *Hausdorff separation axiom*.

Every metric space satisfies the Hausdorff axiom: if ρ is a metric on a space X which generates its topology and $x, y \in X$, $x \neq y$, then the ε-balls $O_\varepsilon(x)$ and $O_\varepsilon(y)$, where $\varepsilon = \frac{1}{2}\rho(x, y)$, are disjoint neighbourhoods of the points x and y.

It is clear that if a space X satisfies an axiom T_i, where $i = 0, 1$ or 2, then each subspace of X satisfies the same axiom.

Proposition 14. *Let $\mathcal{T}_1$ and $\mathcal{T}_2$ be two topologies on a set X such that $\mathcal{T}_1 \subset \mathcal{T}_2$. If the topological space $(X, \mathcal{T}_1)$ satifies axiom T_i where $i = 0, 1$ or 2, then so does the topological space $(X, \mathcal{T}_2)$.*

2.7. Regular and Normal Spaces. The Axioms T_3 and T_4. The separation axioms considered in this section pertain to the separation of points and closed sets from closed sets.

Definition 7. A topological space X is called *regular* if for each closed set $A \subset X$ and any point $x \in X \backslash A$ there exist open sets U and V in X such that $x \in U$, $A \subset V$, and $U \cap V = \varnothing$.

Here is an equivalent condition: for each neighbourhood O_x of an arbitrary point $x \in X$ there exists a neighbourhood O_{1x} of x such that $\bar{O}_{1x} \subset O_x$.

Warning. Regularity of a space does not guarantee that its points are closed: the antidiscrete space is regular. The situation changes if we combine regularity with axiom T_0.

In connection with this we introduce the following axiom.

Axiom T_3. The topological space is regular and satisfies axiom T_0.

It is clear that T_3 implies T_2. A subspace of a regular space is regular and if a space X satisfies axiom T_3, then each subspace also satisfies axiom T_3. A metrizable space is regular: to obtain disjoint neighbourhoods of a point x and a closed set A which does not contain x, we can take the sets $O_\varepsilon(x)$ and $O_\varepsilon(A)$ where $\varepsilon = \frac{1}{2}\rho(x, A) > 0$.

Example 23. Let $X = \mathbb{R}$ be the real line and set $A = \{1/n : n \in \mathbb{N}^+\}$. Call a set closed if and only if it is of the form $B \cup C$ where B is closed in the usual topology $\mathcal{T}$ on $\mathbb{R}$ and $C \subset A$. The complements of the closed sets in X defined in this manner form a topology $\mathcal{T}_1$ on $\mathbb{R}$ for which $\mathcal{T} \subset \mathcal{T}_1$ and $\mathcal{T}_1 \neq \mathcal{T}$ because $\mathbb{R} \backslash A \in \mathcal{T}_1$ and $\mathbb{R} \backslash A \notin \mathcal{T}$. The space $X = (\mathbb{R}, \mathcal{T}_1)$ satisfies the Hausdorff separation axiom (see Proposition 14) but is not regular: A is closed in X, $0 \notin A$, but the point 0 and the set A can not be separated by disjoint neighbourhoods in X

Definition 8. A topological space X is called *normal* if any disjoint closed subsets A and B of X can be separated by disjoint neighbourhoods; that is, there exist open subsets U and V of X such that $A \subset U$, $B \subset V$ and $U \cap V = \varnothing$.

Axiom T_4. The topological space is normal and satisfies axiom T_1.

Axiom T_3 follows from T_4. The class of normal topological spaces is very important and quite broad.

Proposition 15. *Any metrizable space X is normal.*

◁ Let ρ be a metric generating the topology on X, and let A and B be disjoint, nonempty closed subsets of X. For $x \in A$ and $y \in B$ we set $V_x = O_{\varepsilon_x}(x) = \{z \in X : \rho(x, z) < \varepsilon_x\}$ where $\varepsilon_x = \rho(x, B)/3 > 0$ and $U_y = O_{\varepsilon_y}(y) = \{z \in X : \rho(y, z) < \varepsilon_y\}$, where $\varepsilon_y = \rho(y, A)/3 > 0$. The sets $V = \bigcup\{V_x : x \in A\}$ and $U = \bigcup\{U_y : y \in B\}$ are disjoint neighbourhoods of the sets A and B respectively. This follows from the triangle axiom.

Warning. A subspace of a normal T_1 space can fail to be normal; that is, axiom T_4 is not inherited by subspaces. This is one of the main incoveniences in dealing with the class of normal spaces. We will give an example later (see § 5, part 5.6).

Proposition 16. *Every closed subspace of a normal space is normal.*

Example 24 (The Nemytskiĭ plane). Let $L = \{(x, y) \in \mathbb{R}^2 : y \geqslant 0\}$ be the upper half plane and let M be the line $y = 0$. For $z \in M$ and $\varepsilon > 0$, let $W(z, \varepsilon)$ denote the set of points inside the circle in L of radius ε which is tangent to M at z. Set $W_i(z) = W(z, 1/i) \cup \{z\}$ for $i \in \mathbb{N}^+$. If $z \in L \backslash M$, then we take $W_i(z)$ to be the set of all points in L which lie inside the circle of radius $1/i$ with center at the point z, $i \in \mathbb{N}^+$. The set $\mathscr{B} = \{W_i(z) : z \in L, i \in \mathbb{N}^+\}$ is a base of a topology on L. The set L together with this topology is called the *Nemytskiĭ plane*. It is easy to check that L is a regular Hausdorff space. The subset M is closed in the Nemytskiĭ plane; M is discrete as a subspace of L. At the same time, the topology induced on $P = L \backslash M$ by the Nemytskiĭ plane is the same as its topology as a subspace of $\mathbb{R}^2$ with the usual topology. Thus, the Nemytskiĭ plane is a union of two metrizable subspaces M and P, one of which is open and the other closed. However, the Nemytskiĭ plane itself is not metrizable; this follows from Proposition 15 and the assertion below.

Proposition 17. *The Nemytskiĭ plane L is not normal.*

◁ Let C be the set of all points of $P = L \setminus M$ both of whose coordinates are rational. The set C is countable and everywhere dense in L since it intersects every element of the basis $\mathscr{B}$. Suppose that L were a normal space. An arbitrary subset A of M is closed in L as is the set $M \setminus A$. This means that we can choose neighbourhoods U_A of A and V_A of $M \setminus A$ in L. Set $C_A = U_A \cap C$ for $A \subset M$. If $A \subset M, B \subset M$ and $A \neq B$, then $\bar{U}_A \neq \bar{U}_B$. In fact, if $A \setminus B \neq \varnothing$, then $\bar{U}_A \cap V_B \supset U_A \cap V_B \supset A \setminus B \neq \varnothing$ whereas $\bar{U}_B \cap V_B = \varnothing$. But $\bar{U}_A = \overline{U_A \cap C} = \bar{C}_A$ since C is everywhere dense in L and U is open. Consequently, since $A \subset M, B \subset M$ and $A \neq B$, it follows that $C_A \neq C_B$ (otherwise, we would have $\bar{U}_A = \bar{U}_B$). But the cardinality of the set of all countable subsets of a countable set C is equal to $2^{\aleph_0}$ and the cardinality of the set of all subsets of the line M is greater than the cardinality of M; that is, is greater than $2^{\aleph_0}$. This is a contradiction; hence the space is not normal. ▷

Thus, not every regular space in which all finite sets are closed is normal.

The topology of the Nemytskiĭ plane contains the topology generated on the half plane by the usual metric. Thus, Proposition 14 does not generalize to axiom T_4: a topology which is stronger than a normal Hausdorff topology need not be normal (see, also, Example 23).

Proposition 18. *If X is a Hausdorff space in which all but finitely many points are isolated, then X is normal.*

We also cite another broad class of normal spaces.

Proposition 19. *Every subspace of a linearly ordered space is normal.*

It is clear that normality of X is equivalent to the following condition: for each closed subset A of X and any neighbourhood U of A, there exists an open set V such that $A \subset V \subset \bar{V} \subset U$. This condition has a useful variant pertaining to finite open covers of a space.

A collection γ of subsets of a topological space X is called a *cover* of X if $\bigcup \gamma = X$ (that is, if the union of all subsets which are members of γ is X). A cover δ is called *open* (*closed*) if all its elements are open (closed) sets. We also consider indexed covers $\{U_s : s \in S\}$ where some sets have different indexes. Open covers play a fundamental role in topology; we shall develop this theme in more detail later. Here, we cite only the following "*combinatorial shrinking lemma*".

Proposition 20. *A topological space X is normal if and only if for each finite open cover $\gamma = \{U_1, \ldots, U_k\}$ of the space there exists an open cover $\gamma^1 = \{V_1, \ldots, V_k\}$ of X such that $\bar{V}_i \subset U_i$ for all $i = 1, \ldots, k$. The cover γ^1 is called here a shrinking of the cover γ.*

◁ Let A and B be disjoint closed subsets of X. Then $U_1 = X \setminus A$ and $U_2 = X \setminus B$ are open sets which together cover X. Two sets F_1 and F_2 with $F_1 \subset U_1, F_2 \subset U_2$ are closed and such that $F_1 \cup F_2 = X$ if and only if $X \setminus F_1$ and $X \setminus F_2$ are disjoint open neighbourhoods of A and B. If F_i is closed with $F_i \subset U_i$ and if X is normal, then there exist open sets V_i such that $F_i \subset V_i \subset \overline{V}_i \subset U_i$. This is the argument in the case $k = 2$. The rest of the argument proceeds in the usual way by induction. ▷

The following property of normal spaces, the so-called "*combinatorial thickening lemma*" is often applied in dimension theory.

Proposition 21. *Let X be a normal space and $\{F_1, \ldots, F_k\}$ a finite collection of closed subsets of X. Then there exists a collection $\{U_1, \ldots, U_k\}$ of open subsets of X such that $F_i \subset U_i$ for $i = 1, \ldots, k$ and $F_{i_1} \cap \cdots \cap F_{i_p} = \varnothing \Leftrightarrow U_{i_1} \cap \cdots \cap U_{i_p} = \varnothing$ for any $i_1, \ldots, i_p \in \{1, \ldots, k\}$.*

As has already been mentioned, any metrizable space possesses "good" separation properties; in particular, it is normal. Hence, it is little wonder that the separation axioms are crucially involved in the formulation of sufficient conditions for metrizability. A classical result in this direction is due to Uryson: every normal T_1 space with a countable base is metrizable. Tikhonov showed that normality could be replaced by regularity in the latter assertion (see [T1], [ArP]).

A general metrizability criterion is cited in [ArP]. There are several different ways to produce examples of nonmetrizable normal spaces. One can take any linearly ordered nonmetrizable space (see Proposition 19) – in particular, the space of all ordinal numbers, not exceeding the first uncountable ordinal. Another nonmetrizable normal space is the Sorgenfrey line (see § 1, Example 6). It is worth noting that both these spaces are hereditarily normal.

In conclusion, we remark that a topological space satisfying a separation axiom T_i is usually called a T_i-space.

§ 3. Continuous Maps of Topological Spaces: The Foundations of the Theory

The basic means of comparing sets endowed with some additional structure are mappings which preserve the relations between sets and points connected with the structure. In particular, one uses this approach to establish when such objects (consisting of different points, in general) should be considered the same.

The prinicipal means of comparing topological spaces are continuous maps.

3.1. Different Definitions of Continuity of Maps of Topological Spaces. The definition of continuity of a map between topological spaces which most fully corresponds to intuition is that formulated in terms of the nearness relation between points and sets.

Definition 1. Let $(X, \mathcal{T}_1)$ and $(Y, \mathcal{T}_2)$ be topological spaces. A map $f: X \to Y$ is called a *continuous map* of the space X to the space Y if, whenever a point $x \in X$ is near to a set $A \subset X$ in $(X, \mathcal{T}_1)$, the point $\{f(x)$ is near to the set $f(A)$ in $(Y, \mathcal{T}_2)$; that is, if $x\delta A \Rightarrow f(x)\delta f(A)$.

A map between topological spaces which is not continuous is called *discontinuous*.

Proposition 1. *The following conditions are equivalent to the condition that a map f from a topological space X to a topological space Y be continuous*:
(a) *the preimage $f^{-1}(V)$ of every open set V in Y is open in X*;
(b) *the preimage $f^{-1}(P)$ of any closed set P in Y is closed in X.*

The characterizations of continuous maps given in Proposition 1 are technically very simple and convenient although they are somewhat less intuitive than Definition 1. It frequently turns out to be convenient to work with the following characterization of continuity modelled on the well known "ε-δ" definition of continuity of real functions of a real variable.

Proposition 2. *A map f of a topological space X to a topological space Y is continuous if and only if for each point $x \in X$ and each neighbourhood V of f(x) in Y, there exists a neighbourhood U of x in X such that $f(U) \subset V$.*

It is clear that a map is continuous if and only if the image of the closure of any set is contained in the closure of the image of the set: $f(\overline{A}) \subset \overline{f(A)}$ for all $A \subset X$. The first definitions of continuity pertained to maps between subsets of euclidean space and, later, to maps between metric spaces, and were formulated either in the language of ε-δ neighbourhoods or in terms of sequences.

Definition 2. Let (X, ρ_1) and (Y, ρ_2) be metric spaces. A map $f: X \rightarrow Y$ is said to be a *continuous map of the metric space* (X, ρ_1) to the metric space (Y, ρ_2) if for each point $x \in X$ and each positive number ε there exists a positive number δ such that, whenever $x' \in X$ and $\rho_1(x', x) < \delta$, it follows that $\rho_2(f(x'), f(x)) < \varepsilon$.

It follows from Proposition 2 that a map of metric spaces (X, ρ_1) and (Y, ρ_2) is continuous if and only if it is continuous as a map of the topological spaces $(X, \mathcal{T}_{\rho_1})$ and $(Y, \mathcal{T}_{\rho_2})$.

Definition 3. A map $f: X \rightarrow Y$ between topological spaces X and Y is called *sequentially continuous* if, for every sequence $\{x_i\}_{i=1}^{\infty}$ in X converging to a point $x \in X$, the image $\{f(x_i)\}_{i=1}^{\infty}$ converges to the point $f(x)$.

It is obvious that a continuous map of topological spaces is sequentially continuous. The converse is not true.

Example 1. Let A be an uncountable set, $b \notin A$ a point, and $X = A \cup \{b\}$. Call a set $U \subset X$ open if $b \notin U$ or if $A \backslash U$ is countable. This defines a topology on X. The topological space X is not discrete: the point b is not isolated in X. However the intersection of any countable family of open sets in X is open. A space with this property is called a *P-space*. As in any *P*-space, all countable sets in X are closed. Thus, the space X is not sequential – it simply has no nontrivial convergent sequences. The identity map of the space X to the discrete space $Y = A \cup \{b\}$ is not continuous since b is near A in X, but far from A in Y. But the map is sequentially continuous since each convergent sequence in X is constant after some term.

Proposition 3. *If $f: X \rightarrow Y$ is a sequentially continuous map between topological spaces X and Y and if the space X is sequential, then f is a continuous map.*

In particular, for metrizable spaces, continuity is equivalent to sequential continuity. It is precisely this principle which explains why the classical approach to defining continuity using convergent sequences was adequate at first: at that time, all considerations occured within the framework of metrizable spaces.

Definition 4. A continuous map $f: X \to Y$ of topological spaces X and Y is called a *condensation* if f is one to one and onto; that is, if $f(X) = Y$ and $f(x_1) \neq f(x_2)$ whenever $x_1, x_2 \in X$ are such that $x_1 \neq x_2$.

Condensations occur, in particular, when considering two different but comparable topologies on the same set. If $\mathcal{T}_1$ and $\mathcal{T}_2$ are topologies on a set X and $\mathcal{T}_1 \subset \mathcal{T}_2$, then the identity map $I: X \to X$ (where $i(x) = x$ for all $x \in X$) is a condensation of the space $(X, \mathcal{T}_2)$ onto the space $(X, \mathcal{T}_1)$.

We now give a definition of the fundamental notion of a *homeomorphism*.

Definition 5. A map $f: X \to Y$ of topological spaces X and Y is called a *homeomorphism* if f is one to one and onto and satisfies the condition that, for all $x \in X$ and $A \subset X$, $x\delta A \Leftrightarrow f(x)\delta f(A)$.

Proposition 4. *A map $f: X \to Y$ is a homeomorphism of topological spaces X and Y if and only if f is one to one and onto and both f and its inverse f^{-1} are continuous; that is, both are condensations. In this case, f^{-1} is also a homeomorphism.*

Homeomorphisms are characterised as the one to one, onto maps for which the images and preimages of open sets are open, and the images and preimages of closed sets are closed.

Proposition 5. *Any topological space Y is the image under a condensation of a discrete space X.*

We can take X to be the set of points of Y endowed with the discrete topology and consider the identity map $i: X \to Y$.

Thus not every condensation is a homeomorphism.

Definition 6. Two topological spaces X and Y are called *homeomorphic* if there exists a homeomorphism $f: X \to Y$ between them.

The identity map of a space to itself is a homeomorphism, the inverse of a homeomorphism is a homeomorphism, and the composition of homeomorphisms is a homeomorphism. It follows that the relation of being homeomorphic is reflexive, symmetric and transitive; that is, it is an equivalence relation.

If X and Y are homeomorphic spaces, then every property of the space X which can be expressed purely in terms of the nearness relation on X, that is, in terms of the topology of X, is also a property of the topological space Y. In this sense, the topological spaces X and Y are identical.

Definition 7. Let Φ be some property which a topological space can have. If, whenever X possesses property Φ, it follows that every space homeomorphic to X possesses Φ, then Φ is called a *topological property* or a *topological invariant*.

This definition makes precise the intuitively clear, but formally somewhat hazy, notion of a property of a space which can be expressed in terms of its topology.

The overriding "major" problem is to learn to recognize whether or not two topological spaces are homeomorphic. Unfortunately, this problem has no solution for any wide class of spaces: no one has succeeded in finding a finite collection of "computable" (by any kind of algorithm) collection of topological invariants, whose coincidence would guarantee that spaces are homeomorphic. To prove that two topological spaces are not homeomorphic, it suffices to find a topological property present in one space and not in the other. For this purpose, it is desirable to have the widest possible spectrum of topological invariants, mutually independent and as different in nature as possible.

Example 2. Any two disks on the plane are homeomorphic; a disk is homeomorphic to the inside of a square or triangle – the corresponding homeomorphisms are easy to construct directly. An interval $(a, b) = \{x \in \mathbb{R}: a < x < b\}$ on the real line $\mathbb{R}$ is homeomorphic to $\mathbb{R}$. From this it is evident that boundedness of a set and or its diameter are not topological invariants. This is not surprising: boundedness and diameter are definied in terms of a metric, and not in terms of a collection of open sets. A circle is not homeomorphic to a segment $[a, b] = \{x \in \mathbb{R}: a \leqslant x \leqslant b\}$ because any continuous map of a segment of itself has a fixed point, while a rotation of a circle through $90°$ about its center has no fixed point. The property that every continuous map of a topological space to itself have a fixed point is an example of a nontrivial topological invariant. Since every continuous real function on the interval $[a, b] \subset \mathbb{R}$ is bounded, the interval $[a, b]$ is not homeomorphic to the real line $\mathbb{R}$ or, as a result, to the interval (a, b). Although there exists a one to one map of the line $\mathbb{R}$ onto the plane $\mathbb{R}^2$ (see [A6], [ArP, Chapter III]), the spaces $\mathbb{R}$ and $\mathbb{R}^2$ are not homeomorphic – this easily follows from connectedness considerations (see §8). With the help of more refined considerations connected with dimension, it can be shown that when $n \neq m$, n-dimensional euclidean space $\mathbb{R}^n$ is not homeomorphic to m-dimensional euclidean space $\mathbb{R}^m$. Even if a homeomorphism between spaces X and Y exists, this is not always obvious.

Example 3. Every countable metrizable space without isolated points is homeomorphic to the space of rational numbers with the usual topology (see [A6], [E]). The usual space J of irrational numbers is homeomorphic to the Baire space $B(\aleph_0)$ (see §2, example 10). A Cantor perfect set is homeomorphic to the topological product of countably many copies of the discrete two point space $D = \{0, 1\}$ (see §7).

One can have two different topologies $\mathscr{T}_1$ and $\mathscr{T}_2$ on a set X, while the spaces $(X, \mathscr{T}_1)$ and $(X, \mathscr{T}_2)$ are homeomorphic. In connection with this, we mention that the Sorgenfrey line is not homeomorphic to the line $\mathbb{R}$ with the usual topology: the first space is not metrizable, while the second is. Here metrizability is the distinguishing topological invariant.

Other topological invariants include separability, all separation axioms, connectedness, compactness, dimension and many other properties considered in this and other articles.

Important topological invariants of an algebraic nature (homology groups, homotopy groups, and so forth) are constructed and studied within the framework of algebraic topology (see [A5], [P1]). However, the latter are most effective for studying a very narrow range of topological spaces – notably, manifolds and polyhedra.

3.2. General Aspects of the Comparison of Topological Spaces. As mentioned in the preceeding section, there is no effective procedure for distinguishing topological spaces up to homeomorphism. In connection with this, one might phrase the comparison problem for topological spaces more broadly.

General Question 1. How can one determine whether a topological space X can be continuously mapped onto a topological space Y?

It is well known that a segment can be continuously mapped onto a square (the Peano curve, see [A6], [Ar3], and Part II of this book) and that a Cantor perfect set admits a continuous map onto any closed bounded subset of euclidean space $\mathbb{R}^n$ (see [A2], [ArP], Chap. III).

Within the framework set up by Question 1, we single out the following question.

Question 2. Suppose that we are given a "standard" topological space X (for example, X might be a segment, interval, Cantor perfect set, the space of irrational numbers, the Sorgenfrey line, and so forth). What kind of topological invariants characterize the topological spaces onto which X can be continuously mapped?

It is known that the Hausdorff spaces which are continuous images of the Cantor set are precisely the metric compacta and that the Hausdorff spaces which are continuous images of a segment are precisely the locally connected, metrizable, connected compacta (see [Ku1], [ArP]).

Questions 1 and 2 lead to the general question.

Question 3. What kind of topological properties of a space are inherited by its image under a continuous map.

In connection with Question 3, one should keep in mind that every topological space is a continuous image[1] of a discrete space (see Section 3.1). Hence, neither metrizability nor normality is preserved under continuous maps – even if one prescribes rather extensive restrictions on the image (requiring, for example, that it be Hausdorff or a T_3-space).

The individual comparison problem for two arbitrary given topological spaces, namely, the problem of deciding whether it is possible to continuously map one space to another, turns out to be insoluble in principle. It is necessary to replace the problem with a broader sort of problem: that of comparing entire classes of

---

[1] We shall use the phrase *continuous image* to mean the "image under a continuous map".

topological spaces (by means of mappings). Problems 1, 2, and 3 can also be replaced by problems at this level of generality.

General Problem 4. Let $\mathscr{P}$ and $\mathscr{Q}$ be classes of topological spaces. When can each space in class $\mathscr{Q}$ be represented as the continuous image of some space in class $\mathscr{P}$?

This problem is close in spirit to the following.

Problem 5. When does there exist a universal object in a class $\mathscr{P}$, universal being understood to be with respect to continuous maps of objects. That is, when does there exist a space $X \in \mathscr{P}$ such that each space $Y \in \mathscr{P}$ is a continuous image of X?

It is also natural to impose restrictions on the maps used to compare topological spaces; that is, to fix a class of maps. This gives rise to interesting variants of the questions above.

We have already mentioned one special class of continuous maps: the condensations. Here are variants of Question 4 pertaining to condensations.

Problem 6. Let $\mathscr{P}$ and $\mathscr{Q}$ be two classes of topological spaces.

(a) When is each space in $\mathscr{Q}$ the image under a condensation of some space in $\mathscr{P}$?

(b) When can each space in $\mathscr{Q}$ be condensed onto some space in class $\mathscr{P}$?

Much attention in general topology has been directed at the following questions: What kind of spaces condense onto metric spaces? What kind of spaces condense onto compacta? The answer to the first question is known, but nontrivial; there is no complete answer to the second question. For more about this see Chap. IV of [ArP].

3.3. Imbedding Problems. Imbedding problems play an important role in general topology.

Definition 8. A continuous map $f: X \to Y$ which is a homeomorphism of the space X onto the subspace $f(X)$ of Y is called an *imbedding* of the space X into Y.

Here is one of the main variants of the imbedding problem.

Problem 7. Let $\mathscr{P}$ be a given class of topological spaces. Determine (if possible) "standard" topological spaces Y into which it is possible to imbed each space in $\mathscr{P}$. Find the simplest such space Y.

A variant of Problem 7 is the following *universal object problem*:

Problem 8. Which classes $\mathscr{P}$ of topological spaces contain a space X into which each space in $\mathscr{P}$ imbeds?

Some fundamental results relating to Problems 7 and 8 have been obtained: *Tikhonov's imbedding theorem* for Tikhonov spaces to the Tikhonov cube (see 7), the imbedding theorem for arbitrary inductively zero-dimensional Hausdorff spaces into a topological product of two point sets, the *Nobeling-Pontryagin imbedding theorem* (see [AP], [HW]) for n-dimensional metrizable compacta into $(2n + 1)$-dimensional euclidean space, Alexandrov's imbedding theorem for arbitrary T_0-spaces into a power of the connected two point space, and many other results which will be dealt with later. The significance of imbedding theorems is that they allow one to represent abstract, axiomatically presented topological spaces as subspaces of standard, "concrete" topological spaces.

We cite here three of the most elementary (but, nevertheless, very important) results of this type.

Theorem 1 (see [U], [T1]). *Every regular T_1-space with a countable base can be imbedded in the Hilbert space H.*

Thus the Hilbert space H is universal with respect to imbeddings for the class of all T_3-spaces with countable base.

Theorem 2 (see [Ar5]). *A topological space X imbeds in the Cantor perfect set (in the space of irrational numbers) if and only if X has a countable base of open-closed sets and is a T_1-space.*

Spaces with a base of open-closed sets are called *inductively zero-dimensional spaces*. Examples include discrete spaces, the spaces of rational and irrational numbers, the Sorgenfrey line, and the Baire space $B(\tau)$.

Theorem 3 (See [A6], [E]). *A metric space X imbeds in the Baire space $B(\tau)$ (see § 1, example 10) if and only if X possesses a base $\mathscr{B}$ of open-closed sets where the cardinality of $\mathscr{B}$ is less than or equal to τ.*

3.4. Open Mappings and Closed Mappings. Here are two of the simplest principles for classifying maps of topological spaces.

1) Impose restrictions on the behaviour of open and closed sets under the mapping.

2) Impose restrictions on preimages of points and sets or on what happens to subspaces under the mapping.

Definition 9. A map f of a topological space X to a topological space Y is called *open* if the image of each open subset of X under f is an open subset of Y.

Definition 10. A map f of a space X to a space Y is called *closed* if the image of each closed subset of X is closed in Y.

The composition of open maps is an open map, and the composition of closed maps is a closed map.

A homeomorphism is both an open map and a closed map.

Proposition 6. *Let X and Y be topological spaces and let $f: X \to Y$ be a one to one and onto map (in particular, $f(X) = Y$). Then the following conditions are equivalent:* a) *f is a homeomorphism;* b) *f is a continuous open map;* c) *f is a continuous closed map.*

Warning. The condition that a map be open is not equivalent to the condition that it be closed, even when the map is continuous. This stems from the fact that the image of the complement of a set need not be the complement of its image.

Example 4. The projection π of a plane $\Pi = \mathbb{R}^2$ onto a line l which it contains parallel to another line l_1 meeting l is a continuous open map with respect to the usual topologies on $\mathbb{R}^2$ and l.

Let P be the branch of a hyperbola in Π, one of whose asymptotes is l_1. The set P is closed in Π, but its image on l under projection parallel to l_1 is not closed: the point at which l_1 and l intersect is near, but not contained in, the set $\pi(P)$. Consequently, $\pi: \Pi \to l$ is a continuous map which is open but not closed.

The projection of a topological product onto any factor is always a continuous open map (see § 7).

Example 5. Every continuous real function on the segment $I = [a, b] \subset \mathbb{R}$ with the usual topology is a closed map of the segment to $\mathbb{R}$. Moreover, every continuous map of I to a metrizable space is closed – this easily follows from the compactness of I (see § 5). On the other hand, "bending" an interval onto a circle or a "figure eight" gives a continuous closed map which is not open.

The concepts of open and closed maps, being very natural, lead to important variants of Problems 1–5.

Problem 9. Let $\mathscr{P}$ be some class of topological spaces. Characterize the class of all spaces Y which are imges of spaces in class $\mathscr{P}$ under continuous open maps.

Usually Problem 9 is considered under the supplementary assumption that Y satisfy a separation axiom.

An example of a classical result in the spirit of Problem 9 is V.I. Ponomarev's theorem [Po], as generalized by Michael [M1]:

Theorem 4. *Spaces satisfying the first axiom of countability, and only such, are images of metrizable spaces under continuous open maps.*

It is appropriate to contrast Theorem 4, which has found applications in mathematical logic (see the appendix to [Ku2]) to the following result due to Junnila.

Theorem 5. *Any T_1-space Y is a continuous open image* (that is, the image under a continuous open map) *of a hereditarily normal space X in which each point is the intersection of countably many open sets.*

The image of a metrizable space under a closed map may fail to satisfy the first axiom of countability and may even not be metrizable (see 3.5); these spaces were characterized by N.S. Lashnev [L].

Theorem 6 ([St2]). *If a T_1-space Y is the image of a metric space X under a continuous map which is both open and closed, then Y is metrizable.*

Any continuous map of a compact space to a Hausdorff space is closed. This attests to the importance of the concept of a closed map.

Example 6. Any map $f: W \to \mathbb{C}$ of a domain W in the complex plane to the complex plane $\mathbb{C}$ which is analytic at each point of W is a continuous open map (see [Sto], [Wh]).

This fact can be used to prove the "fundamental theorem of algebra" which asserts that any polynomial of positive degree has a complex root.

In order to address questions like Problems 1–5, it is important to have as wide as possible a spectrum of topological properties preserved by open (closed) maps.

It is easy to show that the image under a continuous open map of a space satisfying the first axiom of countability is again a space satisfying the first axiom of countability. Open maps carry spaces with a countable base to spaces with a countable base.

Proposition 7. *If X is a Frechet-Uryson space and $f: X \rightarrow Y$ is a continuous closed map with $f(X) = Y$, then Y is also a Frechet-Uryson space.*

◁ Let $B \subset Y$, $y \in Y$ and $y \in \bar{B}$. Set $A = f^{-1}(B)$. The set $f(\bar{A})$ is closed in Y and $B = f(A) \subset f(\bar{A})$. Consequently, $y \in f(\bar{A})$. Choose a point $x \in \bar{A}$ for which $f(x) = y$ and choose in A a sequence $\{x_n\}_{n=1}^{\infty}$ converging to x. Then the sequence $\{y_n\}_{n=1}^{\infty}$ where $y_n = f(x_n) \in f(A) = B$ converges to y. ▷

Proposition 7 implies the following theorem.

Theorem 7 ([St2]). *The image of a metrizable space under a continuous closed map is a Frechet-Uryson space.*

Note that the image above need not satify the first axiom of countability.

3.5. Restrictions on the Preimages of Points Under Maps. Restrictions on the preimages of points are among the more elementary restrictions that can be imposed on a map.

A map $f: X \rightarrow Y$ is *finite to one* (or has *finite multiplicity*) if the preimage $f^{-1}(y)$ of any point $y \in Y$ consists of a finite number of points; a map is said to be *countable to one* (or to have *countable multiplicity*) if all preimages of points are countable. If the subspace $f^{-1}(y)$ of X has a countable base for all $y \in Y$, the map $f: X \rightarrow Y$ is called an *s-map*.

Placing restrictions on the preimages of points allows us to enrich our possibilities for classifying spaces by means of maps in an essential way. In particular, we have the following.

Theorem 8 (Ponomarev [Po]). *A T_1-space Y is the image of a metrizable space under a continuous open s-map if and only if Y possesses a point-countable basis.*

In this connection, a base $\mathcal{B}$ of Y is said to be *point-countable* if, for each point $y \in Y$, the set of elements of $\mathcal{B}$ containing y is countable.

In the same spirit as Theorem 8, there is a characterization of images of metrizable spaces under continuous open maps in which the preimages of points are compact (Arkhangel'skiĭ [Ar1]).

In algebraic topology, the theory of Riemann surfaces and the theory of functions of a complex variable, the consideration of universal covers naturally gives rise to maps in which the preimages of points are discrete. A map $f: X \rightarrow Y$

is called *discrete* if the preimage $f^{-1}(y)$ of each point $y \in Y$ is a closed discrete subspace of the space X. In applications, discrete maps are most often encountered in conjunction with local homeomorphisms.

A map $f: X \to Y$ of topological spaces X and Y is called a *local homeomorphism* if each $x \in X$ has an open neighbourhood U in X such that f homeomorphically maps the subspace U onto an open subspace $f(U)$ of Y.

Local homeomorphisms are discrete, open, continuous maps.

Open finite-to-one maps of two dimensional manifolds play a role in describing analytic functions on manifolds (see Whyburn [Wh] and Stoilow [Sto]).

Example 7. An analytic map $f: W \to \mathbb{C}$ of a domain W in the complex plane $\mathbb{C}$ to the complex plane $\mathbb{C}$ whose derivative is not equal to zero at any point is a local homeomorphism.

A map is called *monotone* if the preimage of any point is connected (see §8) and zero-dimensional if the preimage of any point is zero-dimensional (see 1). Continuous open zero-dimensional maps are said to be *light*.

In geometric topology an important place is occupied by questions involving the possibility of mapping Euclidean cubes of lower dimension onto Euclidean cubes of higher dimension by an open and zero-dimensional or open and monotone mapping (Keldysh [K], Anderson – see §8 and [AF] and [AP]).

3.6. Unions of Maps and Conditions that They be Continuous. Let X be a space covered by a collection $\{X_\alpha: \alpha \in A\}$ of subspaces and suppose, that, for each $\alpha \in A$, we are given a map $f_\alpha: X_\alpha \to Y$ of the space X_α to Y (where Y is the same for every $\alpha \in A$). Suppose, in addition, that if $x \in X_{\alpha'} \cap X_{\alpha''}$ for α', $\alpha'' \in A$, then $f_{\alpha'}(x) = f_{\alpha''}(x)$; let us agree to call a collection of maps $\{f_\alpha: \alpha \in A\}$ satisfying this condition *compatible*. The maps f_α can then be "glued together" into a single "global" map $f: X \to Y$ defined by the rule $f(x) = f_\alpha(x)$ when $x \in X_\alpha$. The map f is called the *union* of the maps f_α and we write $f = \bigcup_{\alpha \in A} f_\alpha$. If is natural to ask what conditions guarantee that f be continuous. With the notation above, we have the following results.

Proposition 8. *If each subspace X_α is open in X for all $\alpha \in A$ and if all the maps f_α are continuous, then the union $f: X \to Y$ is continuous. If, in addition, every f_α is open, or a local homeomorphism, then so is f.*

Proposition 9. *If A is a finite set, if X_α is closed in X for each $\alpha \in A$ and if all the maps f_α are continuous, then the map f is continuous. If, in addition, every f_α is closed, then so is $f: X \to Y$.*

The difference in the formulations of Propositions 8 and 9 is connected with the fact that the union of any collection of open sets is open whereas the analogous assertion for closed sets is false.

It is clear that even a union of only two continuous maps might be discontinuous.

3.7. Decomposition Spaces. In mathematical descriptions of how one geometric figure is obtained from another, one frequently encounters the intuitively appealing operation of "gluing" or "identification". "Gluing" the endpoints of a segment together gives a topological model of a circle; "identifying" all boundary points of a disk with a single point gives a model of a sphere; gluing corresponding points on opposite sides of a square to one another gives a torus; gluing one side of a square to the opposite side by identifying "skew-symmetric" points (that is, pairs of points symmetric with respect to the center of the square) gives a Möbius strip; identifying diametrically opposite points of a two dimensional sphere gives a projective plane, and so forth.

The examples enumerated above and many other constructions of this sort have a simple and natural description within the confines of general topology.

A *decomposition* of a topological space X is a collection γ of pairwise disjoint nonempty subsets of X which covers X; that is, $\bigcup \gamma = X$.

To each subset A of X which is not a single point, we associate a decomposition $\gamma_A = \gamma(X, A)$ of X called the *A-decomposition*, the only element of which is not a single point is A: $\gamma_A = \{\{A\}, \{x\}: x \in X \backslash A\}$.

The topology $\mathcal{T}$ on a space X induces a topology $\mathcal{T}^* = \mathcal{T}|_\gamma$ on a decomposition γ called the *natural* (or *canonical*) *topology* of the decomposition γ. Namely, a subset $V \subset \gamma$ belongs to the topology $\mathcal{T}|_\gamma$ if the union $\bigcup V$ of all elements of γ belonging to V (as subsets of X) is open in X; that is, if $\bigcup V \in \mathcal{T}$. It is easy to check that the collection $\mathcal{T}|_\gamma$ of subsets of γ is, in fact, a topology on γ. The topological space $(\gamma, \mathcal{T}|_\gamma)$ is called the *decomposition space γ* of the space X.

Subsets of a space X which are unions of a subcollection of γ are called *distinguished* (or *γ-distinguished*). The collection $\langle \mathcal{T} \rangle_\gamma$ of all open distinguished subsets of X is in canonical one-to-one correspondence with the collection of all open subsets of the decomposition space γ. This allows us to analyze the properties of the space γ using the collection $\langle \mathcal{T} \rangle_\gamma$. This is convenient because the elements of the latter are more intuitively accessible. An element of $\langle \mathcal{T} \rangle_\gamma$ is called a representative of the corresponding set in the decomposition space.

Example 8. If $X = [0, 1]$ is the usual unit segment and $A = \{0, 1\}$, then the space of the standard decomposition γ_A is homeomorphic to the circle. If A is the boundary of a disk X in Euclidean space, then the space of the standard decomposition γ_A of X is homeomorphic to the two dimensional sphere S^2.

Example 9. Let X be the Euclidean plane $\mathbb{R}^2$ and A a line in it. Consider the space of the standard decomposition γ_A of the space X. The collection $E = \langle T \rangle_{\gamma_A}$ of all distinguished subsets of X consists of the open sets of X containing A or not intersecting A. In the decomposition space γ_A, the point A is an intersection of a countable collection of open sets "represented" by bands of width $1/n$, $n \in \mathbb{N}^+$, containing the line A. However, the space γ_A does not satisfy the first axiom of countability at A. This is easy to prove by a standard Cantor type "diagonal" argument. The space γ_A does not have a countable base.

Thus, passing to the simplest standard decomposition space can take one out of class of spaces with a countable base, even if we begin with a space as "classical" as the plane $\mathbb{R}^2$.

It can be shown that the space γ_A is Hausdorff, normal and a Frechet-Uryson space. General assertions from which this follows will be quoted below.

3.8. Quotient Spaces and the Quotient Topology. Closely connected with the concept of a decomposition space is the concept of a quotient space and a quotient topology.

Let $(X, \mathcal{T})$ be a topological space and suppose that f is a map of X to some set Y on which no topology has been specified. What topology can one put on Y to best reflect the properties of the topology $\mathcal{T}$ on X and the map f? It is reasonable to take a topology $\mathcal{T}'$ on Y with respect to which the map $f: X \to Y$ becomes continuous. However, this requirement can be trivially satisfied by taking $\mathcal{T}'$ to be the antidiscrete topology on Y. It is easy to see that there can be other topologies on Y which make the map $f: X \to Y$ continuous. In order to single out a unique topology on the set Y associated to a map f and a topology $\mathcal{T}$ on X, we take the strongest topology on Y which makes f continuous. The existence of such a topology is immediate.

Definition 11. The *quotient topology* on a set Y (generated by a map $f: X \to Y$ and a topology $\mathcal{T}$ on X) is the collection $\mathcal{T}|_f = \{V \subset Y: f^{-1}(V) \in \mathcal{T}\}$ of all subsets of Y whose preimages are open in X.

Since the preimage of an intersection of sets is equal to the intersection of the preimages and the preimage of a union is equal to the union of the preimages, the collection $\mathcal{T}|_f$ is, in fact, a topology on the set Y.

Proposition 10. *The topology $\mathcal{T}|_f$ is the strongest topology $\mathcal{T}'$ for which the map f is continuous (that is, $\mathcal{T}' \subset \mathcal{T}|_f$ for any topology $\mathcal{T}'$ which makes f continuous).*

Warning. At first glance, the following approach to defining a canonical topology $\mathcal{T}^*$ on Y given a map f and topology $\mathcal{T}$ seems no less natural: take $\mathcal{T}^*$ to be the topology on Y generated by the subbase $\{f(U): U \in \mathcal{T}\}$ of all images of open subsets of X. It is obvious that f becomes open with this choice of topology $\mathcal{T}^*$ on Y; however, f need not be continuous.

Example 10. Let $X = [0, 1] \cup [2, 3]$ be the subspace of the line $\mathbb{R}$ with the usual topology and let Y be the subspace of the plane $\mathbb{R}^2$ formed by two perpendicular segments intersecting in their middles (a "cross"). Map the segment $[0, 1]$ linearly to one of the segments of the "cross" Y and $[2, 3]$ linearly to the other segment of Y. The images of the sets $[0, 1]$ and $[2, 3]$, which are open in X, are sets V_1 and V_2 which intersect at the midpoint C of the cross. Since $V_1 \cap V_2 = \{C\}$ the one point set $\{C\}$ must be in the topology $\mathcal{T}^*$. But the preimage of the set $\{C\}$ under f is the two point set $U = \{\frac{1}{2}, 2\frac{1}{2}\}$. Since U is not an open set, the map f of X to $(Y, \mathcal{T}^*)$ is not continuous.

Using the example of f above, one can show that, given a map $f: X \to Y$ and a topology $\mathcal{T}$ on X, it is not always possible to define a topology $\mathcal{T}'$ on Y so that the map f becomes both continuous and open (or continuous and closed).

The following definition is closely related to Definition 11.

Definition 12. A map f of a topological space $(X, \mathscr{T})$ onto a topological space $(Y, \mathscr{T}_1)$ is said to be a *quotient map* if $V \in \mathscr{T}_1 \Leftrightarrow f^{-1}(V) \in \mathscr{T}$; that is, if a set is open in Y if and only if its preimage is open in X.

A quotient map which is one to one is a homeomorphism.

The connection between Definitions 11 and 12 is expressed by the following assertion.

Proposition 11. *A map f of a topological space $(X, \mathscr{T})$ onto a topological space $(Y, \mathscr{T}_1)$ is a quotient map if and only if the topology $\mathscr{T}_1$ coincides with the quotient topology $\mathscr{T}|_f$ on the set Y generated by the map f and the topology $\mathscr{T}$ in accord with Definition 11; that is, if $\mathscr{T}_1 = \mathscr{T}|_f$.*

Any quotient map is continuous.

Proposition 12. *Any continuous open map of a topological space X onto a topological space Y is a quotient map.*

Proposition 13. *Let f be a continuous map of a space X onto a space Y. Then the following conditions are equivalent:* a) f *is a quotient map;* b) *if the preimage* $f^{-1}(B)$ *of a set* $B \subset Y$ *is closed in X, then B is closed in Y.*

Proposition 14. *Any continuous closed map of a topological space X onto a topological space Y is a quotient map.*

The fact that the set of quotient maps includes both the class of all continuous open maps and the class of all continuous closed maps underlines its importance.

Definition 11 allows us to easily generate examples of quotient spaces. However, much more enlightening examples arise in connection with decomposition spaces.

3.9. Quotient Spaces and Decomposition Spaces. Let γ be a decomposition of a topological space $(X, \mathscr{T})$. To each point $x \in X$ we associate the unique element $\pi_\gamma(x) = A$ of γ containing x. This defines a canonical map $\pi_\gamma\colon X \to \gamma$ of X to γ, the so-called *projection map*.

Proposition 15. *The quotient topology $\mathscr{T}|_{\pi_\gamma}$ on γ generated by the map π_γ and the topology $\mathscr{T}$ coincides with the canonical topology $\mathscr{T}|_\gamma$ on the decomposition γ and the canonical map $\pi_\gamma\colon X \to Y$ of X to the decomposition space γ is a quotient map.*

By choosing an arbitrary topological space and any decomposition γ of it, we obtain, by Proposition 15, a large supply of examples of quotient maps.

Any map f of a topological space X onto a topological space Y generates a decomposition γ of X whose elements are preimages of points under the map $f\colon \gamma(f) = \{f^{-1}(y)\colon y \in Y\}$. Give the decomposition $\gamma(f)$ the canonical decomposition topology and denote the resulting space by $X_f = X|_f$. Let φ be the standard map between X_f and Y: if $A = f^{-1}(y) \in \gamma(f)$, then $\varphi(A) = y$. It is clear that $\varphi\colon X_f \to Y$ is one to one. The relation between the space X_f and the space Y is described in terms of φ as follows.

Proposition 16. *The map φ is continuous if and only if f is continuous (in which case φ is a condensation).*

Proposition 17. *The map f is a quotient map if and only if φ is a homeomorphism.*

In other words a quotient map can be characterized as a map whose image is canonically homeomorphic to the decomposition space it generates.

Propositions 15 and 17 justify using the term "quotient space" to refer to a decomposition space of a topological space (and not just to the image of a topological space under a quotient map).

In the preceding section, we remarked that the class of quotient maps includes all continuous open maps onto and all continuous closed maps onto. This gives rise to the following natural question: which decompositions correspond to open mappings and which to closed mappings?

Definition 13. A decomposition γ of a topological space $(X, \mathscr{T})$ is *upper continuous* if, for each element A of the decomposition γ and each open subset U of X containing A, there exists a distinguished open subset V of X such that $A \subset V \subset U$.

Proposition 18. *The canonical map $\pi_\gamma \colon X \to \gamma$ of a space X to its decomposition space γ is closed if and only if the decomposition γ is upper continuous.*

Example 11. The space of a standard decomposition γ_A of a topological space X where $\gamma_A = \{\{A\}, \{x\} \colon x \in X \backslash A\}$ and A is closed is upper continuous. In particular, if A is a straight line on the Euclidean plane $X = \mathbb{R}^2$, then the canonical map $\pi \colon X \to \gamma_A$ of the plane to the space of the decomposition γ_A is closed (and continuous). As remarked in Example 10, the space γ_A is not metrizable (it does not satisfy the first axiom of countability). Consequently, neither metrizability nor the first axiom of countability need be preserved by a continuous closed map to a normal T_1-space.

A decomposition γ of a space X is upper continuous if and only if, for each open subset U of X, the union of all elements of γ contained in U is open in X.

Definition 14. A decomposition γ of a space X is *lower continuous*, if for each open subset U of X, the union of all elements of γ intersecting U is open in X.

Proposition 19. *A decomposition γ of a space X is lower continuous if and only if the canonical map $\pi_\gamma \colon X \to \gamma$ of X to the decomposition space γ is open.*

It is now easy to construct an example of a quotient map which is neither open nor closed: one takes a decomposition of a topological space which is neither upper nor lower continuous.

Example 12. Let $X = \mathbb{R}$ be the usual space of real numbers and let $A_n = \{1/n, n\}$ for $n \in \mathbb{N}^+$. Set $\gamma = \{A_n\}_{n=1}^\infty \cup \{\{x\} \colon x \in X \backslash \bigcup_{n=1}^\infty A_n\}$. Thus, γ is a decomposition consisting of one point and two point sets. The decomposition γ is not upper continuous: the interval $U = (-1, 1)$ is an open subset of X and contains the element $\{0\}$ of the decomposition γ, but there is no distinguished open subset which contains $\{0\}$ and lies in U (see Definition 13). Nor is the decomposition γ

lower continuous: the union of all elements of γ which intersects the interval $U = (-1, 1)$ is the set $W = (-1, 1) \cup \{n, 1/n; n \in \mathbb{N}^+\}$ which is not open in the space $X = \mathbb{R}$.

Definition 15. A decomposition of a topological space is called *completely continuous* if it is both upper and lower continuous.

From Propositions 18 and 19, we get the following.

Proposition 20. *A decomposition γ of a topological space X is completely continuous if the canonical map $\pi_\gamma: X \to \gamma$ of X to the decomposition space γ is both closed and open.*

Decomposition spaces (more precisely, upper continuous decomposition spaces) were first considered by Alexandrov and Hopf in [AH].

3.10. Attaching One Space to Another by a Map.

The concept of a quotient space can be used to describe the following construction which plays an important role in topology, especially algebraic topology.

Let X and Y be disjoint topological spaces and $f: A \to Y$ a continuous map of a closed subset A of X to Y. Consider the decomposition on the free sum $Z = X \oplus Y$ whose elements are $\gamma = \{\{f(a)\} \cup f^{-1}f(a): a \in A\} \cup \{\{x\}: x \in X \setminus A\} \cup \{\{y\}: y \in Y \setminus f(A)\}$.

We denote the decomposition space γ by $X \cup_f Y$ and say that it is obtained by *attaching* the space X to Y by the map f. The map f is called the *attaching map*.

Example 13. Let X and Y be any nonempty topological spaces and $I = [0, 1]$ the unit segment. Attach the product $X \times I \times Y$ to the free sum $X \oplus Y$ of X and Y by the map f defined on the set A of all points of the form $(x, 0, y)$ and $(x, 1, y)$ which takes $(x, 0, y)$ to x and $(x, 1, y)$ to y. The resulting space is denoted $X * Y$ and called the *join* of the spaces X and Y: it consists of the points of the spaces X and Y connected by segments running from any point $x \in X$ to any point $y \in Y$ where none of the segments have common points except the endpoints.

Proposition 21. *Let X, Y and $f: A \to Y$ be as at the beginning of this section. Let $p: X \oplus Y \to X \cup_f Y$ be the natural map. Then we have:* a) *p maps the space Y homeomorphically to the closed subspace $p(Y)$ of $X \cup_f Y$;* b) *p maps the subspace $X \setminus A$ homeomorphically onto an open subspace of $X \cup_f Y$.*

3.11. Separation Axioms in Decomposition Spaces.

The concept of a decomposition space is a very effective and transparent means of constructing diverse examples of topological spaces. However, in practice it is often important to be able to guarantee that examples satisfy "good" separation axioms.

Separation axioms are very easily lost by passing to decomposition spaces unless one imposes strong additional restrictions on the decompositions.

Proposition 22. *A decomposition space γ of a space X satisfies the T_1 separation axiom if and only if every element of the decomposition γ is a closed subset of X.*

Example 14. Let $X = I^2 = \{(x, y): 0 \leqslant x \leqslant 1, 0 \leqslant y \leqslant 1\}$ be the unit square, $\eta = \{\{(0, y)\}: 0 \leqslant y \leqslant 1\}$ the collection of all one point subsets lying on the left vertical side of I^2 and $\xi = \{I_x: 0 < x \leqslant 1\}$ where each $I_x = \{(x, y): 0 \leqslant y \leqslant 1\}$ is a vertical segment. The partition $\gamma = \eta \cup \xi$ of X is lower continuous, but not upper continuous. All elements of γ are closed in X so that the decomposition space γ is a T_1-space. The space X is metrizable and, a fortiori, normal and Hausdorff. However, the decomposition space γ does not satisfy the Hausdorff separation axiom: the points $A = \{(0, 0)\}$ and $B = \{(0, 1)\}$ of γ are not separated by disjoint neighbourhoods since every open γ-distinguished set in X containing one of the points $(0, 0)$ and $(0, 1)$ certainly contains an entire ε-strip: namely, the set $\bigcup\{I_x: 0 < x < \varepsilon\}$ for some $\varepsilon > 0$.

The situation is fundamentally different for upper continuous decomposition.

Proposition 23. *If γ is an upper continuous decomposition of a normal space X, then the decomposition space γ is also normal.*

◁ Let $\pi: X \to \gamma$ be the canonical map and let F_1 and F_2 be disjoint closed sets in γ. The sets $P_1 = \pi^{-1}(F_1)$ and $P_2 = \pi^{-1}(F_2)$ are closed in X and do not intersect; consequently, there exist open subsets U_1 and U_2 of X such that $P_1 \subset U_1$, $P_2 \subset U_2$ and $U_1 \cap U_2 = \varnothing$. Since the decomposition γ is upper continuous, there exist distinguished open subsets W_1 and W_2 of X for which $P_1 \subset W_1 \subset U_1$ and $P_2 \subset W_2 \subset U_2$. The sets $V_1 = \pi(W_1)$ and $V_2 = \pi(W_2)$ are open in the space γ, $V_1 \cap V_2 = \varnothing$ (since $\pi^{-1}(V_1) \cap \pi^{-1}(V_2) = W_1 \cap W_2 = \varnothing$) and $F_1 \subset V_1$, $F_2 \subset V_2$. ▷

Theorem 9. *The image of a normal T_1-space under a closed continuous map is a normal T_1-space.*

In particular, the image of a metrizable space under a closed continuous map is a normal T_1-space. This is not the case for continuous open maps (see Example 14 and Proposition 19).

3.12. Restricting Maps to Subspaces. Let X and Y be topological spaces and $f: X \to Y$ a map. If X_1 is a subspace of X, let $f_1 = f|X_1$ denote the *restriction of the map f* to X_1; that is, the map $f_1: X_1 \to Y$ of the topological space X_1 to the topological space Y defined by the rule: $f_1(x) = f(x)$ for all $x \in X_1$. It is clear that if f is continuous, then f_1 is also continuous. However, openness, closedness, and being a quotient map are not stable under the operation of restriction.

Example 15. Let f be orthogonal projection of the plane $\mathbb{R}^2$ onto a straight line Y contained in it. Set $X_1 = Y \cup Y_1$ where Y_1 is the straight line in X perpendicular to Y. The map $f: X \to Y$ is continuous and open, but the restriction $f_1: X_1 \to Y$ to the closed subspace X_1 of X is not open: the set $U = X_1 \backslash Y$ is open in X_1 and is carried by f_1 onto the one-point set $Y \cap Y_1$ which is not open in Y.

Example 16. We "wind" a segment $X = [a, b]$ (where $b > a + 1$) twice around a circle Y. The resulting map $f: X \to Y$ is continuous and closed. The restriction f_1 of f to the open subspace $X_1 = (a, b)$ of X is not a closed map: for sufficiently large $k \in \mathbb{N}^+$, the closed subset $P_k = \{a + 1/n: n \in \mathbb{N}^+, n \geqslant k\}$ of X_1 is mapped onto a set which is not closed in Y. Note that here $Y = f_1(X_1)$.

Proposition 24. *Let f, X, Y, and X_1 be as at the beginning of this section. Then a) if $f: X \to Y$ is a closed map and X_1 is closed in X, then $f_1: X_1 \to Y$ is a closed map, b) if $f: X \to Y$ is an open map and X_1 is open in X, then $f_1: X_1 \to Y$ is an open map.*

Proposition 25. *Let X and Y be topological spaces, $f: X \to Y$ a map, Y_1 a subspace of Y and $X_1 = f^{-1}(Y_1)$. Let $f_{11}: X_1 \to Y_1$ be the restriction to the total preimage, that is, $f_{11}(x) = f(x)$ for all $x \in X_1$. If f is an open (resp., closed) map, then f_{11} is also an open (resp., closed) map.*

◁ Suppose that U_1 is open in X_1. There exists an open subset U of X such that $U \cap X_1 = U_1$. Then $f(U) \cap Y_1 = f_{11}(U_1)$ since $X_1 = f^{-1}(Y_1)$ and $f(U \setminus X_1) \cap Y_1 = \varnothing$. But $f(U)$ is open in Y. Hence $f_{11}(U_1)$ is open in Y. ▷

Warning. Keeping the notation of Proposition 25, it is not true, in general, that if $f: X \to Y$ is a quotient map, then the double restriction to the total preimage $f|_{Y_1}: X_1 \to Y_1$ is a quotient map. In the language of decomposition spaces, this circumstance can be expressed as follows. Let γ be a decomposition space of X and let X_1 be a distinguished subset of X; that is, $X_1 = \bigcup \gamma_1$ for some subcollection γ_1 of γ. Then γ_1 is a decomposition of the subspace X_1 of X and carries a canonical topology $\mathcal{T}_1$ as a decomposition space. On the other hand, γ_1 carries a topology $\mathcal{T}_1'$ as a subspace of the decomposition space γ. These two topologies on γ_1 need not coincide: in general, the former is stronger than the latter (that is, $\mathcal{T}_1' \subset \mathcal{T}_1$). In fact, the intersection of a γ-distinguished open subset of X with X_1 is a γ_1-distinguished open set in the space X_1, but not every open γ_1-distinguished subset of X_1 is an intersection with X_1 of an open γ-distinguished subset of X. In order to see this, we turn to Example 12. The set $X_1 = X \setminus \bigcup_{n=1}^{\infty} A_n$ is the union of all single point elements of the decomposition γ; consequently, it is distinguished. The topology on the decomposition space $\gamma_1 = \{\{x\}: x \in X_1\}$ of X_1 can be identified via the correspondence $x \to \{x\}$ with the topology on X_1. In particular, the point $\{0\} \in \gamma_1$ is far from the set $P = \{\{x\}: x > 1, x \notin \mathbb{N}^+\}$. But in the subspace γ_1 of the decomposition space γ the point $\{0\}$ is near P, because it is near it in the space γ. We can express this as follows: the canonical map $\pi: X \to \gamma$ is a quotient map, but its restriction to the total preimage $\pi_{11} = \pi|_{\gamma_1} = \pi|_{\gamma_1}: X_1 \to \gamma_1$ is not a quotient map because it is one to one but not a homeomorphism.

Proposition 26. *Let X and Y be topological spaces, $f: X \to Y$ a quotient map and Y_1 an open or closed subspace of Y. Then the restriction map $f|_{Y_1}: X_1 \to Y_1$ to the total preimage $X_1 = f^{-1}(Y_1)$ of Y_1 is a quotient map.*

3.13. Some Properties of Quotient Maps and Hereditarily Quotient Maps

Proposition 27. *A continuous map f of a topological space X to a topological space Y is a quotient map if and only if, for each non-closed set $B \subset Y$, there* *exists a point* $y \in \bar{B} \backslash B$ *such that* $f^{-1}(y) \cap \overline{f^{-1}(B)} = \varnothing$. *Equivalently, there exists a point* $x \in \overline{f^{-1}(B)}$ *for which* $f(x) \notin B$.

Theorem 10 (see [Fr1]). *The image of a sequential space under a quotient map is a sequential space.*

◁ Let $f: X \to Y$ be a quotient map, $f(X) = Y$, $B \subset Y$, and suppose that B is not closed in Y. Since the space X is sequential, there exists a point $x \in \overline{f^{-1}(B)} \backslash f^{-1}(B)$ to which some sequence $\{x_n\}_{n=1}^{\infty}$ of points of $f^{-1}(B)$ converges. Since f is continuous, the sequence $\{f(x_n)\}_{n=1}^{\infty}$ of points of B converges to a point $y = f(x)$. It is clear that $y \in \bar{B} \backslash B$.

Warning. The image of a separable metrizable space under a quotient map may fail to be a Frechet-Uryson space – one need only consider the map in Example 12.

Definition 16. A map f of a topological space X onto a topological space Y is said to be a *hereditarily quotient map* if, for each subspace $Y_1 \subset Y$, the double restriction of f to the total preimage $X_1 = f^{-1}(Y_1)$ (that is, the map $f|_{Y_1}: X_1 \to Y_1$) is a quotient map.

In Section 3.12 we saw that not every quotient map is a hereditarily quotient map. But it follows from Proposition 25 that continuous open maps onto and continuous closed maps onto are hereditarily quotient maps.

Proposition 27 results from the following characterization of hereditarily quotient maps.

Proposition 28. *A continuous map f of a space X onto a space Y is a hereditarily quotient map if and only if $f^{-1}(y) \cap \overline{f^{-1}(B)} \neq \varnothing$ whenever $y \in \bar{B}$ with $y \in Y$ and $B \subset Y$.*

Proposition 28 implies the following.

Theorem 11 ([Ar2]). *The image of a Frechet-Uryson space under a hereditarily quotient map is a Frechet-Uryson space.*

The following criterion for a map to be a hereditarily quotient map is sometimes convenient.

Theorem 12 (Arkhangel'skiĭ [Ar2]). *A continuous map f of a space X onto a space Y is a hereditarily quotient map if and only if, for each point $y \in Y$ and each open subset U of X containing the preimage $f^{-1}(y)$ of y, the interior Int $f(U)$ of $f(U)$ contains y (that is, $f(U)$ is a neighbourhood of y in Y).*

In view of the characterization of a hereditarily quotient map given by Theorem 12, such maps are called *pseudo-open* (see [Ar2]).

Theorem 11 implies the following.

Corollary 1. *The image of a metrizable space under a pseudo-open map is a Frechet-Uryson space.*

As shown earlier (see Example 9), such images need not satisfy the first axiom of countability. The following fact is worth noting in connection with Theorems 11 and 12.

Theorem 13. *Every quotient map of a topological space to a Hausdorff Frechet-Uryson space is a hereditarily quotient map (that is, is pseudo-open).*

3.14. The Canonical Quotient Map Associated to an Intersection of Topologies. Let $\{\mathscr{T}_s : s \in S\}$ be a collection of topologies on a set X. Set $X_s = X \times \{s\} = \{(x, s) : x \in X\}$ and endow X_s with the topology $\mathscr{T}_{s,s} = \{U \times \{s\} : U \in \mathscr{T}_s\}$. The map $i_s \colon X_s \to X$ defined by $i_s(x, s) = x$ is a homeomorphism of the space $(X_s, \mathscr{T}_{s,s})$ to the space $(X, \mathscr{T}_s)$.

Let $\mathscr{T}$ denote the intersection of the topologies $\mathscr{T}_s$: that is, $\mathscr{T} = \bigcap \{\mathscr{T}_s : s \in S\}$. The free sum $\tilde{X} = \sum_{\oplus} X_s$ of the topological spaces X_s is mapped canonically to $(X, \mathscr{T})$ by the map $p \colon \tilde{X} \to X$ whose restriction to each X_s coincides with i_s (that is, $p(x, s) = x$ for all $(x, s) \in \tilde{X}$.

The space X_s is naturally identified with the space $(X, \mathscr{T}_s)$ (and the topology $\mathscr{T}_s$ with the topology $\mathscr{T}_{s,s}$). In this sense, the space $\tilde{X}$ is the free sum of the topological spaces $(X, \mathscr{T}_s)$ over $s \in S$.

Proposition 29. *The canonical map $p \colon \sum_{\oplus} X_s \to (X, \mathscr{T})$ of the free sum of the spaces $(X, \mathscr{T}_s)$ to the space $(X, \mathscr{T})$, where $\mathscr{T} = \bigcap \{\mathscr{T}_s : s \in S\}$, is a quotient map.*

Proposition 29 and Theorem 11 imply the following.

Theorem 14 (see [Ar4]). *Let $\{\mathscr{T}_s : s \in S\}$ be any collection of metrizable topologies on a set X and let $\mathscr{T} = \bigcap \{\mathscr{T}_s : s \in S\}$ be their intersection. Then $(X, \mathscr{T})$ is a sequential T_1-space. Conversely, the topology of any sequential T_1-space is an intersection of a collection of metrizable topologies.*

Definition 17 (see [Ar4]). A collection $\{\mathscr{T}_s : s \in S\}$ of topologies on a set X is said to be *compatible* if the canonical map $p \colon \sum_{\oplus} X_s \to (X, \mathscr{T})$ where $\mathscr{T} = \bigcap \{\mathscr{T}_s : s \in S\}$ is a hereditarily quotient (that is, pseudo-open) map.

It is clear from Definition 17 that two topologies $\mathscr{T}_1$ and $\mathscr{T}_2$ on a set X are compatible if and only if, for any sets $U_1 \in \mathscr{T}_1$ and $U_2 \in \mathscr{T}_2$, the set $U_1 \cup U_2$ is a neighbourhood of every point of $U_1 \cap U_2$ in the topological space $(X, \mathscr{T}_1 \cap \mathscr{T}_2)$.

Theorem 15. *The topology of a space X is an intersection of a compatible collection of metrizable topologies if and only if X is a Frechet-Uryson T_1-space.*

3.15. Functional Separation; Complete Regularity and Tikhonov Spaces. At first blush, the idea of "separating" points and closed sets of a topological space by continuous, real-valued functions must seem somewhat specialized in view of the high degree of abstraction inherent in the notion of a topological space. The real numbers are not even mentioned in the axioms of a topological space. Thus, the wide range of topological spaces satisfying such a separation axiom involving continuous real-valued functions seems astonishing. This section is devoted to such spaces.

The following separation axiom opens the way for the deep penetration of the real numbers into general topology. This is exceedingly important. For instance, it is along these lines that general topology is tightly connected with the theories of linear topological spaces and topological rings.

Definition 18. A topological space X is said to be *completely regular* if, for every set $A \subset X$ and any point $x \in X \backslash A$ which is far from A (that is, not belonging to the closure of A), there exists a continuous function $f \colon X \to \mathbb{R}$ on X such that the point $f(x)$ is far from the set $f(A)$; that is, $f(x) \notin \overline{f(A)}$. (If the latter holds, we say that the function f *separates the point x and the set A*).

The following characterization of complete regularity is often used as a definition.

Proposition 30. *A topological space X is completely regular if, for each point $x_0 \in X$ and for each closed set $A \subset X$ which does not contain x_0, there exists a real-valued continuous function f on X such that $f(x_0) = 0$ and $f(x) = 1$ for all $x \in A$.*

It is easy to verify that every completely regular space is regular and that every completely regular T_0-space satisfies the Hausdorff separation axiom. It is not so easy to construct an example of a regular space which is not completely regular (see [My], [Nv]).

We mention a broad class of completely regular spaces.

Proposition 31. *If X is an inductively zero-dimensional topological space (that is, if X has a base consisting of open-closed sets), then X is completely regular.*

◁ It suffices to remark that if U is an open and closed subset of X, then the function $f \colon X \to \mathbb{R}$ defined by setting $f(x) = 0$ for $x \in U$ and $f(x) = 1$ for $x \in X \backslash U$ is continuous. ▷

The following fact (*Uryson's lemma*) is of fundamental significance for general topology and its applications: every normal topological space is completely regular (see §6).

A completely regular T_1-space is called a *Tikhonov space* (after Tikhonov who introduced them into topology and proved the fundamental theorem about embedding Tikhonov spaces into the Tikhonov cube I^τ (see §7).

The notion of functional separation also appears in the following more specialized, but still important, concept.

Definition 19. A topological space X is said to be *functionally Hausdorff* if, for any two different points x, y of X there exists a continuous function $f \colon X \to \mathbb{R}$ such that $f(x) \neq f(y)$.

Every functionally Hausdorff space satisfies the Hausdorff separation axiom, but the converse does not hold. The following characterization of functionally Hausdorff spaces is useful.

Proposition 32. *A space X is functionally Hausdorff if and only if it admits a condensation to a Tikhonov space.*

3.16. The Weak Topology Generated by a Collection of Maps. Let X be a set without a prescribed topology, and suppose that we are given a collection of maps f_α of X into topological spaces $(Y_\alpha, \mathcal{T}_\alpha)$ where $\alpha \in A$. If we give X the discrete topology, then all the maps f_α become continuous. But there is a more economical way to define a topology on X which makes all the maps f_α continuous: take $\mathcal{T}$ to be the topology generated by the subbase $\mathcal{P} = \{ f_\alpha^{-1}(U) : U \in \mathcal{T}, \alpha \in A \}$. In this case we say that the topology $\mathcal{T}$ on X is *generated* by the collection of maps $\{ f_\alpha : \alpha \in A \}$. If $\mathcal{T}'$ is any topology on X for which every map f_α, $\alpha \in A$, is continuous, then it is clear that $\mathcal{T} \subset \mathcal{T}'$.

This construction plays a fundamental role in defining the topology of the product of topological spaces and in defining the topology on the limit of an inverse spectrum of topological spaces (see §7 and §9).

It is also closely connected with the concept of complete regularity: a topological space X is completely regular if and only if its topology is generated (in the above sense) by the collection of all continuous real-valued functions on X.

Example 17. Let X be a functionally Hausdorff space and $\mathcal{F}$ the collection of all continuous real-valued functions on X. Then X endowed with the topology generated by the family of maps $\mathcal{F}$ is a Tikhonov space onto which the space X condenses (see Proposition 32).

Proposition 33. *A topological space X is inductively zero-dimensional if and only if its topology is generated by the collection of all continuous maps of X to the discrete two point space $D = \{0, 1\}$.*

§4. Some Metric Properties and their Relationship with Topology

4.1. Maximal ε-Discrete Subspaces and ε-Dense Sets. A subset $A \subset X$ of a metric space (X, ρ) is said to be ε-*dense* in (X, ρ) (where ε is a fixed positive number) if $\rho(x, A) < \varepsilon$ for all $x \in X$; that is, if for every $x \in X$, there exists a point $x' \in A$ for which $\rho(x, x') < \varepsilon$.

A subset $A \subset X$ of a metric space (X, ρ) is called ε-*discrete* in (X, ρ) if $\rho(x, x') \geqslant \varepsilon$ for every $x, x' \in A$ such that $x \neq x'$.

The collection $\mathcal{P}$ of all ε-discrete subsets of a metric space (X, ρ) is ordered by inclusion and, hence, there exists a maximal chain C in $\mathcal{P}$. The union $A^* = \bigcup C$ is a maximal ε-discrete set in X; that is, if we add another point of X to A^*, the set is no longer ε-discrete. Consequently, A^* is ε-dense in (X, ρ). This gives the following result.

Proposition 1. *If (X, ρ) is any metric space and $\varepsilon > 0$, then there exists a subset which is both ε-discrete and ε-dense.*

If A is an ε-discrete subset of a metric space (X, ρ), then every subset B of A is closed in the space $(X, \mathcal{T}_\rho)$ and, hence, A is a discrete subspace of $(X, \mathcal{T}_\rho)$. If, in addition, $(X, \mathcal{T}_\rho)$ has a countable base, then A has a countable base and is, therefore, countable.

Proposition 2. *Let (X, ρ) be a metric space. Then the topological space $(X, \mathcal{T}_\rho)$ has a countable base if and only if, for every $\varepsilon > 0$, any ε-discrete subset of (X, ρ) is countable.*

Definition 1. A metric space (X, ρ) is called *totally bounded* if, for each $\varepsilon > 0$, there exists a finite ε-dense subset of (X, ρ); in this case, the metric ρ is also called totally bounded.

Proposition 2 can be used to prove the following.

Proposition 3. *A metric space (X, ρ) is totally bounded if and only if, for every $\varepsilon > 0$, any ε-discrete subset of (X, ρ) is finite.*

Any subspace (Y, ρ) of a totally bounded metric space (X, ρ) is totally bounded.

Example 1. Setting the distance between any two distinct points of an infinite set X equal to one gives a metric space which is not totally bounded (see Proposition 3). Any bounded subset of the line or plane with the usual Euclidean metric is a totally bounded metric space. The real line $\mathbb{R}$ with the usual metric is not totally bounded.

Theorem 1. *A subspace (X, ρ) of a finite-dimensional euclidean space $(\mathbb{R}^n, \rho)$ (see §2, Example 5) is totally bounded if and only if it is bounded in $\mathbb{R}^n$; that is, if and only if there exists a number $a \in \mathbb{R}$ such that $\rho(x, y) < a$ for all $x, y \in X$.*

It is natural to ask which spaces can be metrized by a totally bounded metric? The answer is distinguished by its simplicity and generality.

Theorem 2. *A metrizable space can be metrized by a completely bounded metric if and only if it has a countable base (equivalently, if and only if it is separable.)*

The significance of the notion of totally bounded is clear both from Theorem 1 and from the fact (see §5) that all compact metric spaces are totally bounded.

4.2. Complete Metric Spaces. A sequence $\{x_i\}_{i=1}^\infty$ of points in a metric space (X, ρ) is caled a *Cauchy sequence* or a *fundamental sequence* in (X, ρ) if, for every $\varepsilon > 0$ there exists a natural number k such that $\rho(x_i, x_m) < \varepsilon$ for all $i \geqslant k$ and all $m \geqslant k$. It is obvious that every convergent sequence in a metric space is a Cauchy sequence. The converse is not true: a sequence $\{r_n\}_{n=1}^\infty$ of rational numbers which converges to an irrational number does not converge in the space of rational numbers.

Definition 2. A metric space (X, ρ) is called *complete* if every Cauchy sequence in (X, ρ) converges to a point of X. In this case the metric ρ is said to be *complete*.

Metric spaces which are not complete frequently arise upon passing to subspaces. The following assertion underlines this.

Proposition 4. *Let (X, ρ) be a metric space and (Y, ρ) (more precisely, $(Y, \rho|_Y)$) a subspace. If the metric space (Y, ρ) is complete, then the set Y is closed in $(X, \mathcal{T}_\rho)$.*

It is obvious that a closed subset of a complete metric space is itself a complete metric space.

Example 2. A discrete space with the trivial metric is complete. A Baire space, the Hilbert space H, euclidean space $\mathbb{R}^n$ with the euclidean metric, the space of continuous real functions on the interval $[a, b]$ (see §2, Examples 5, 6, 10) are complete metric spaces. The space J of irrational numbers with the usual metric is an incomplete metric space. However, the Baire space $B(\aleph_0)$ of countable weight is homeomorphic to the space of irrational numbers. Thus, a complete metric space can be homeomorphic to an incomplete metric space; that is, completeness of metric spaces is not a topological invariant.

On the other hand, not every metrizable space can be metrized by a complete metric: there is no complete metric generating the usual topology on the space of rational numbers. This can be deduced from the following assertion.

If (X, ρ) is a complete metric space and if there are no isolated points in the space $(X, \mathcal{T}_\rho)$, then the set X is uncountable; in fact, $|X| \geqslant 2^{\aleph_0}$.

The proof of this assertion is based on the Baire property of complete metric spaces (see Section 4.4).

In connection with Proposition 4, note that it is not necessary that a subspace A of a metrizable space X be closed in order that A can be metrized by a complete metric. For example, the set of irrational numbers is not closed in $\mathbb{R}$. Nevertheless, metrizability by a complete metric imposes an important constraint on the disposition of a subspace in the ambient topological space.

Theorem 3 (see [A1]). *Let Y be an everywhere dense subspace of a Hausdorff space X. If Y is metrizable by a complete metric, then Y is an intersection of a countable collection of open dense subsets of X.*

Criteria for a space to be metrizable by a complete metric are given, for example, in [ArP].

4.3. Completeness and the Extension of Continuous Maps. Recall that a *set of type G_δ* or, more simply, a *G_δ set* in a topological space X is a subset of X which can be represented as an intersection of a countable collection of open subsets of X.

Proposition 5. *If a topological space Y is metrizable by a complete metric, then every continuous map $f: A \to Y$ defined on an everywhere dense subspace A of a topological space X can be extended to a continuous map $\tilde{f}: \tilde{A} \to Y$ on some G_δ set $\tilde{A} \subset X$ containing A.*

The following theorem is due to Lavrent'ev. It has important applications, particularly to dimension theory (see [E]).

Theorem 4. *Let X and Y be topological spaces which are metrizable by complete metrics, and let $A \subset X$ and $B \subset Y$ be any subspaces. Then every homeomorphism f of A to B extends to a homeomorphism $\tilde{f}$ of some subspace $\tilde{A} \subset X$ to a subspace $\tilde{B} \subset Y$, where $A \subset \tilde{A}, B \subset \tilde{B}$ and $\tilde{A}$ and $\tilde{B}$ are G_δ sets in X and Y, respectively.*

4.4. The Baire Property of Complete Metric Spaces and Sets of the First and Second Category

Definition 3. A topological space X is said to have the *Baire property* if the intersection of any countable collection $\{U_n\}_{n=1}^\infty$ of open, everywhere dense subsets of X is everywhere dense in X: $\overline{\bigcap_{n=1}^\infty U_n} = X$.

The following theorem, due to Hausdorff [H2], expresses one of the main topological properties of complete metric spaces.

Theorem 5. *A topological space X which is metrizable by a complete metric possesses the Baire property.*

Baire proved (in 1898) that the real line has the property mentioned in Theorem 5; the latter result is called the *Baire category theorem.*

Theorem 5 can be used to show that various metric spaces are not metrizable by complete metrics.

Example 3. Let $\mathbb{Q} = \{r_n\}_{n=1}^\infty$ be the space of rational numbers. The sets $U_n = \mathbb{Q} \backslash \{r_n\}$ are open and everywhere dense in $\mathbb{Q}$, but $\bigcap_{n=1}^\infty U_n = \mathbb{Q} \backslash \mathbb{Q} = \varnothing$ is the empty set. Hence the space $\mathbb{Q}$ is not metrizable by a complete metric.

The Baire property for metric spaces is not equivalent to the metrizability of spaces by a complete metric.

Proposition 6. *If some everywhere dense subspace Y of a space X has the Baire property, then X has the Baire property.*

Example 4. Consider the subspace of the euclidean plane $\mathbb{R}^2$ consisting of all points $(x, y) \in \mathbb{R}^2$ for which $y > 0$ and all the points $(x, 0) \in \mathbb{R}^2$ for which x is rational. It follows from Proposition 6 that X possesses the Baire property; at the same time it is not metrizable by a complete metric because the closed subspace $Y = \{(x, 0): x \in \mathbb{Q}\}$ of X is not metrizable by a complete metric (see Example 3).

Definition 4. Let X be a topological space and A a subset of X. If A is a union of a countable collection of nowhere dense subsets of X, then A is said to be a set of the *first category* in X. Otherwise, A is said to be a set of the *second category* in X.

Theorem 5 can be reformulated as follows: if a space X is metrizable by a complete metric and if A is a first category subset of X, then $X \backslash A$ is everywhere dense in X. In particular, no complete metric space X is a union of a countable collection of nowhere dense subsets of X. So, a complete metric space without isolated points cannot be countable.

Theorem 5 has numerous applications in and outside of topology. In particular, in the course of the proof of the following theorem, it is applied to analyze the construction of functions of first Baire class on the real line.

Theorem 6 (Baire). *If a real-valued function f on the line ℝ is the pointwise limit of an everywhere convergent sequence of continuous real-valued functions, then f is continuous at all points except for those belonging to a first category subset of ℝ* (see [Ba1], [Ba2]).

One of the simplest proofs of the existence of a nowhere differentiable, everywhere continuous function on ℝ is based on Theorem 5 (Banach, see [Ox]). Theorem 5 also plays a role in the theory of dynamical systems (see the Poincaré recurrence theorem [Ox]) and in the proof of the embedding theorem for n-dimensional compacta into $(2n + 1)$-dimensional euclidean space (the Nöbeling-Pontryagin theorem, see [HW]).

This list of examples could be extended indefinitely: Theorem 5 is actually one of the most fundamental and frequently used results in_all of mathematics. For more details about it see [HW] and [E].

4.5. The Contraction Mapping Principle. The contraction mapping principle is metric, not topological, in nature and is connected in a very essential way with completeness of metric spaces. It has many applications: for example, it can be used to prove existence and uniqueness of solutions of differential equations and it can be used to find approximate solutions of differential equations.

Definition 5. A map $f: X \to X$ of a metric space (X, ρ) to itself is called a *contraction* (*mapping*) if there exists a real number α where $0 \leqslant \alpha < 1$ such that $\rho(f(x), f(y)) \leqslant \alpha\rho(x, y)$ for all $x, y \in X$.

The contraction mapping principle is the following.

Theorem 7. *A contraction mapping $f: X \to X$ of a nonempty complete metric space to itself has exactly one fixed point; that is, there exists a unique $x \in X$ such that $f(x) = x$.*

◁ The proof of Theorem 7 also gives a method for finding the fixed point by successive approximations. Namely, choose any $x_1 \in X$ and inductively set $x_{n+1} = f(x_n)$. This gives a sequence of points in the space (X, ρ) which is a Cauchy sequence because

$$\rho(x_{n+2}, x_{n+1}) = \rho(f(x_{n+1}), f(x_n)) \leqslant \alpha\rho(x_{n+1}, x_n) \leqslant \alpha^n\rho(x_1, x_2).$$

Since (X, ρ) is complete, the sequence $\{x_n\}_{n=1}^{\infty}$ converges to a point $x \in X$. But the sequence $\{f(x_n)\}_{n=1}^{\infty} = \{x_{n+1}\}_{n=1}^{\infty}$ also converges. Consequently, $x = f(x)$ by the continuity of f. If $f(x) = x$ and $f(y) = y$, then $\rho(x, y) = \rho(f(x), f(y)) \leqslant \alpha\rho(x, y)$ from which it follows that $x = y$.

4.6. Completion of Metric Spaces with Respect to a Metric. Let (X, ρ) be a metric space and $\mathscr{E}$ the set of all Cauchy sequences in (X, ρ) which do not have a limit in (X, ρ).

Call Cauchy sequences $\{x_n\}_{n=1}^{\infty}$ and $\{y_n\}_{n=1}^{\infty}$ in $\mathscr{E}$ *equivalent* if $\lim_{n\to\infty} \rho(x_n, y_n) = 0$. The corresponding equivalence relation $\sim$ on $\mathscr{E}$ is reflexive, transitive and symmetric. Let S be the set of all equivalence classes into which the equivalence relation $\sim$ partitions $\mathscr{E}$. If ξ, $\eta \in S$, then we set $\tilde{\rho}(\xi, \eta) = \lim_{n\to\infty} \rho(x_n, y_n)$ where $\{x_n\}_{n=1}^{\infty} \in \xi$ and $\{y_n\}_{n=1}^{\infty} \in \eta$. If $x \in X$ and $\xi \in S$, then we set $\tilde{\rho}(x, \xi) = \lim_{n\to\infty} \rho(x, x_n)$ where $\{x_n\}_{n=1}^{\infty} \in \xi$. Finally, if x, $y \in X$, then we set $\tilde{\rho}(x, y) = \rho(x, y)$. These conventions (and the metric axioms) define a metric $\tilde{\rho}$ on the set $\tilde{X} = X \cup S$. It is clear that the metric space (X, ρ) is a subspace of the metric space $(\tilde{X}, \tilde{\rho})$ and that any Cauchy sequence $\{x_n\}_{n=1}^{\infty}$ in (x, ρ) which does not have a limit in (X, ρ) converges in $(\tilde{X}, \tilde{\rho})$ to the point $\xi \in S \subset X$ where $\{x_n\}_{n=1}^{\infty} \in \xi$. Hence, $\tilde{\rho}(\xi, X) = 0$ for all $\xi \in S$ and the set X is everywhere dense in the space $(\tilde{X}, \mathscr{T}_{\tilde{\rho}})$. The metric space $(\tilde{X}, \tilde{\rho})$ is complete; this can be deduced from the following assertion.

Proposition 7. *Suppose that (X, ρ) is a metric space and that $Y \subset X$ is an everywhere dense subset of $(X, \mathscr{T}_{\rho})$. If every Cauchy sequence in Y converges in $(X, \mathscr{T}_{\rho})$, then the space (X, ρ) is complete.*

Definition 6. A *completion* of a metric space (X, ρ) (with respect to the metric ρ) is a complete metric space $(\tilde{X}, \tilde{\rho})$ which contains (X, ρ) as an everywhere dense metric subspace.

Recall that an *isometry* of a metric space X onto a metric space Y is a one to one map $f\colon X \to Y$ which preserves distances between points. The argument preceding Definition 6 establishes the existence part of the following theorem. The uniqueness part is obvious because the space (X, ρ) is everywhere dense in its completion.

Theorem 8. *Every metric space (X, ρ) has a completion $(\tilde{X}, \tilde{\rho})$ which is unique up to isometries fixing the points of X.*

Remark. A complete metric space can be considered to be the completion of any of its everywhere dense subspaces. This allows one to exhibit examples of metric spaces and their completions. Thus, the completion of the space of rational numbers with the usual metric is the space $\mathbb{R}$ of real numbers. It is obvious that the completion of a complete metric space is the space itself.

Proposition 8. *If (X, ρ) is a totally bounded metric space, then the completion $(\tilde{X}, \tilde{\rho})$ is a totally bounded metric space.*

The notion of metric completeness and the operation of completion with respect to a metric plays an important role in functional analysis (see [Nv], [E]).

§5. Compact Topological Spaces

The notion of a compact topological space plays an important role in topology and its applications. The prototype of this notion is the following property of a segment (Lebesgue, 1903): any open covering of a segment $[a, b] = \{x \in \mathbb{R}:$

$a \leqslant x \leqslant b\}$ of the real line $\mathbb{R}$ contains a finite subcovering. Borel later established that any closed bounded subset of a finite dimensional euclidean space possesses the same property. Not only did this property appear in a wide variety of contexts, but it turned out that many fundamental principles in different areas of mathematics depended on it (such as the Kreĭn-Milman theorem, the Stone-Weierstrass theorem, the theorem that the unit sphere is compact in the weak topology, and so on, not to mention the classical theorems about continuous functions on a segment).

5.1. Different Definitions of Compactness and Related Properties. Recall that an open cover of a topological space X is a collection γ of open subsets of X such that $\bigcup \gamma = X$. If $\mu \subset \gamma$ and $\bigcup \mu = X$ then the collection μ is called a *subcover* of the cover γ.

Definition 1. A topological space X is said to be *compact* if every open cover γ of X has a finite subcover μ.

Proposition 2. *A discrete space X is compact if and only if X is finite.*

Any antidiscrete space is compact. Any finite space is also compact.
A metric space (X, ρ) is said to be compact if the topological space $(X, \mathcal{T}_\rho)$ is compact.

Theorem 1 (Borel). *A subspace X of a finite-dimensional Euclidean space $\mathbb{R}^n$ is compact if and only if X is closed and bounded in $\mathbb{R}^n$.*

Theorem 1 allows one to exhibit many noncompact spaces. However, the noncompactness of such examples can be established directly. For example, the entire line $\mathbb{R}$ and the interval (a, b) are noncompact. The Hilbert space H and the Baire space $B(\tau)$ are also noncompact – we return to these below. On the other hand, segments, squares, cubes, circles, and spheres are compact spaces.

Example 1. An infinite discrete space is not compact: the cover $\{\{x\}: x \in X\}$ by one point sets has no finite subcover. This example shows (as does the example of the line $\mathbb{R}$) that a complete metric space need not be compact.

The following characterization of compactness is perhaps most appealing to the intuition.

Proposition 2. *A topological space X is compact if and only if the intersection of any centered system of closed subsets of X is non-empty.*

Compactness of a space X is also equivalent to the condition that for any centered collection ξ of subsets of X there exists an accumulation point in X; that is, a point x such that $x \in \bar{P}$ for all $P \in \xi$.

Very closely related to the concept of compactness are the concepts of sequential compactness, countable compactness, and pseudocompactness.

Definition 2. A topological space X is called *sequentially compact* if any (countable) sequence $\{x_n\}_{n=1}^{\infty}$ of points in X contains a subsequence which converges in X.

Warning. Not every compact space is sequentially compact: there exist infinite compact spaces in which there are no nontrivial convergent sequences (for example, the Stone-Čech compactification of an uncountable discrete space – see [ArP], [C1], [Wa], [E]). And not every sequentially compact space is compact.

Example 2. Let X be a set of cardinality $\aleph_1$ (where $\aleph_1$ is the first uncountable cardinal; see [ArP], chap. I) and let $<$ be a well-ordering on X such that, for every $x \in X$, the set $\{y \in X : y < x\}$ of predecessors of x in $(X, <)$ is countable (that is, $<$ is an economical well ordering on X). The ordering $<$ generates the topology $\mathcal{T} = \mathcal{T}_<$ of a linearly ordered space on X. The collection $\mathcal{B}$ of all semi-intervals $(a, b] = \{x \in X : a < x \leqslant b$, where $a, b \in X$ and $a < b\}$ is a base of the topology $\mathcal{T}$ (for each element $x \in X$ of the ordered set $(X, <)$ there is a "next" element in $(X, <)$). The space $(X, \mathcal{T})$ is sequentially compact because the closure of every countable subset of $(X, \mathcal{T})$ is a compact space with a countable base (see Proposition 3 below).

However $(X, \mathcal{T})$ is not compact: the open cover of X by all elements of the base $\mathcal{B}$ has no countable subcover. In fact, if it did, the set X would have to be countable since every member of the base $\mathcal{B}$ is a countable set.

Proposition 3. *Every compact Frechet-Uryson space is sequentially compact.*

In particular, any compact space satisfying the first axiom of countability is sequentially compact.

Definition 3. A topological space X is called *countably compact* if every infinite subset $A \subset X$ has a limit point in X; that is, a point x such that the set $O_x \cap A$ is infinite for every neighbourhood O_x of x.

Every sequentially compact space is countably compact, but the converse is false.

Proposition 4. *Every compact space is countably compact.*

Thus, in contrast to sequential compactness, countable compactness follows from compactness. Proposition 4 results from the following characterization of countable compactness.

Proposition 5. *A topological space X is countably compact if and only if every countable open cover of X has a finite subcover.*

5.2. Fundamental Properties of Compact Spaces. The following result is fundamental.

Theorem 2. *If a topological space Y is the image of a compact space X under a continuous map, then Y is compact.*

Theorems 1 and 2 imply the following classical result in real analysis.

Corollary 1. *Any continuous real valued function on a compact topological space is bounded and attains its minimum and maximum values.*

Compactness is not characterized by the property above – any countably compact space enjoys this property.

Not every subspace of a compact space is compact – it is sufficient to consider a segment and an interval.

Theorem 3. *A closed subspace of a compact space is compact.*

A vitally important fact is that the converse is true in the class of Hausdorff spaces. Namely, the following holds.

Theorem 4. *Every compact subspace of a Hausdorff space X is closed in X.*

The property of being closed is relative: it depends on the ambient space. Theorem 4 allows us to view compactness as an absolute closedness, if we restrict ourselves to the broad family of all Hausdorff spaces. In fact, absolute closedness in the class of regular T_1-spaces characterizes compactness (Aleksandrov and Uryson [AH]):

Theorem 5. *If Y is a regular T_1-space which is closed in every Hausdorff space X containing Y as a subspace, then Y is compact.*

It is clear from Theorem 5 that Theorem 4 does not extend to countably compact spaces. It is also not possible to extend Theorem 4 to T_1-spaces.

Example 3. Let X be an infinite set in which the open sets are the empty set and any set whose complement is finite. The space X is compact (since every nonempty open subset of X covers all of X except for a finite number of points) and satisfies the T_1-separation axiom. Every subspace of X is compact (for the same reason) and every infinite subspace of X is everywhere dense. Consequently, any infinite subset $Y \subset X$ (other than X) is a compact subspace of X which is not closed.

The following fundamental assertion about continuous maps of compact spaces into Hausdorff spaces is based on Theorems 2 and 4.

Theorem 6. *Every continuous map of a compact space to a Hausdorff space is closed.*

The following corollary of Theorem 6 is very useful.

Corollary 2. *Every one to one and onto map (that is, condensation) of a compact space to a Hausdorff space is a homeomorphism.*

Neither Theorem 6 nor Corollary 2 extends to continuous maps of compact spaces to T_1-spaces.

Example 4. Let $I = [0, 1]$ be the unit segment with the usual topology and let X be the space obtained by giving I the topology of Example 3 (that is, the only closed sets of X are the finite sets and the whole space).

Then the identity map $i: I \to X$ is a continuous, one to one map of the compact Hausdorff space I to the T_1-space X in which the Hausdorff separation axiom is not satisfied. Consequently, i is not a homeomorphism; that is, i is not a closed and not an open map.

Moreover, neither Theorem 6 nor Theorem 2 generalize to maps between countably compact spaces and Hausdorff spaces (see [E, chap. III]).

5.3. Compact Metric and Compact Metrizable Spaces. A metric space (X, ρ) is said to be compact if the topological space $(X, \mathcal{T}_\rho)$ is compact.

The following theorem completely characterizes compactness of the space $(X, \mathcal{T}_\rho)$ in metric terms.

Theorem 7 (see [A6]). *A metric space (X, ρ) is compact if and only if it is complete and totally bounded.*

◁ Let (X, ρ) be compact. Then X is everywhere dense in the completion $(\tilde{X}, \tilde{\rho})$ of (X, ρ). By Theorem 4, X is closed in $(\tilde{X}, \tilde{\rho})$. Hence $(X, \rho) = (\tilde{X}, \tilde{\rho})$ is a complete space. Let A be an arbitrary ε-network in X. Then A is closed in (X, ρ) and, therefore, compact. But A is also discrete (see §4). Hence A is a finite set (see Proposition 1). This proves necessity. ▷

The completion of a totally bounded space is a totally bounded space. In addition, total boundedness is preserved under passage to subspaces. From this and from Theorem 7 we obtain the following.

Corollary 3. *A metric space (X, ρ) is totally bounded if and only if its completion $(\tilde{X}, \tilde{\rho})$ (with respect to the metric ρ) is compact.*

Corollary 4. *Any totally bounded metric space is a subspace of a compact metric space.*

It is striking that compactness in the case of metric spaces is characterized by the following two conditions.

Theorem 8 (see [E]). *A metrizable space $(X, \mathcal{T})$ is compact if and only every metric ρ on X generating $\mathcal{T}$ (that is, $\mathcal{T} = \mathcal{T}_\rho$) is complete.*

Theorem 9 (see [E]). *A metrizable space $(X, \mathcal{T})$ is complete if and only if every metric on X which generates the topology $\mathcal{T}$ is totally bounded.*

Thus, compactness of metric spaces can be viewed, on the one hand, as absolute (metric) completeness, and on the other, as absolute total boundedness.

We now consider topological characterizations of compactness of metrizable spaces. There is a rather dramatic simplification in this case: the constraints acquire a countable character.

Theorem 10. *Let X be a metrizable space. Then the following conditions are equivalent:*

a) *X is compact;*

b) *X is sequentially compact (that is, every sequence in X has a convergent subsequence);*

c) *X is countably compact (that is, each infinite set in X has a limit point (in X); each countable open cover of the space X has a finite subcover);*

d) *every decreasing sequence* $\{F_n\}_{n=1}^{\infty}$ *of nonempty closed subsets of X* (*that is, every sequence such that* $F_n \supset F_{n+1}$ *for all n*) *has a nonempty intersection.*

Condition d) is naturally associated with the following characterization of metric completeness: a metric space (X, ρ) is complete if and only if for each sequence $\{F_n\}_{n=1}^{\infty}$ of nonempty closed sets of X such that $F_{n+1} \subset F_n$ for all n and $\lim_{n \to \infty}$ diam $F_n = 0$, the intersection $\bigcap_{n=1}^{\infty} F_n$ is nonempty.

It is clear that the latter condition and condition d) are analogues of the "Nested Interval Principle" (see [A6]).

The list of conditions equivalent to compactness will be augmented by another natural condition in the following section.

5.4. Pseudocompact Spaces

Definition 4. A completely regular T_1-space X on which each continuous real valued function is bounded is said to be *pseudocompact*.

Theorem 11. *A metrizable space is compact if and only of it is pseudocompact.*

It is already evident from the definition itself that pseudocompact spaces should play an essential role in functional analysis. This is actually the case; concerning this, see [Ru] and [E, chap. III]).

The following is easily proved.

Proposition 6. *Every countably compact completely regular T_1-space is pseudocompact.*

One might suspect that countable compactness is equivalent or, at least, closely related to pseudocompactness in the class of completely regular T_1-spaces. In fact, pseudocompactness differs essentially from countable compactness. The class of pseudocompact spaces is immeasurably wider than the class of completely regular, countably compact T_1-spaces. This follows from the next assertion.

Theorem 12 (Noble [No]). *Every completely regular T_1-space can be represented as a closed subspace of a pseudocompact space.*

Thus, nothing need remain of pseudocompactness after passing to closed subspaces (except the separation properties which are part of the definition). On the other hand, countable compactness is inherited by closed subspaces (as is compactness). The extent to which pseudocompactness is broader than countable compactness is also clear from the recently constructed examples of infinite pseudocompact spaces in which every countable set is closed and there are no nontrivial convergent sequences.

5.5. Compactness and Separation Axioms.

An important peculiarity of the separation properties in compact spaces is the coincidence of axioms T_2, T_3, and T_4.

Theorem 13. *Every compact Hausdorff space is normal.*

Example 3 shows that not every compact T_1-space satisfies the Hausdorff separation axiom.

It follows from Theorem 13, for example, that not every Hausdorff space is homeomorphic to a subspace of a compact Hausdorff space. In fact, one has the following (see §6 and §7).

Corollary 5. *Every subspace of a compact Hausdorff space is completely regular.*

Theorems 13 and 4, as well as Theorem 6 and Corollary 2 show that compactness takes on strong new properties in the presence of the Hausdorff separation axiom and that the two stronger separation axioms add nothing new. This is one of the reasons why compact Hausdorff spaces have proved to be one of the main objects of investigation in general topology. They even have a short name: a compact Hausdorff space is called a *compactum*.

Warning. Theorem 13 does not extend to countably compact Hausdorff spaces – such spaces need not even be regular.

5.6. Compactifications of Topological Spaces. The most natural analog of (metric) completion in the context of arbitrary topolological spaces is the notion of compactification, which consists in extending a space to a compact space. When and in what ways can a given topological space be represented as a subspace of a compact space? Does there exist a canonical such extension?

In this section, we only consider the most basic concepts and facts pertaining to this large area of general topology.

Definition 5. A compact space cX is called a *compactification* of a topological space X if X is a subspace of cX (in particular, $X \subset cX$) which is everywhere dense in cX.

Let X be a nonempty topological space and ξ any object not contained in X. Set $aX = X \cup \{\xi\}$ and define a topology on aX by declaring the open sets of aX to be the entire space aX and the sets which are open in X. It is clear that aX is compact, $\bar{X} = aX$ and the space X is a subspace of aX. Thus, any space can be extended to a compact space by adding a single point. However, the space aX does not satisfy the T_1 separation axiom – one point subsets of X are not closed in aX. On the other hand, aX satisfies the T_0 separation axiom if X does. The investigation and construction of compactifications becomes deeper and more useful when additional restrictions on separation properties are imposed.

Which T_i-spaces have a compactification which is a T_i-space? Not every T_2 or T_3 space has such a compactification because any subspace of a compact Hausdorff space is completely regular (see Corollary 5).

Theorem 14. *Every completely regular T_1-space has a Hausdorff compactification.*[2]

---

[2] A *compact Hausdorff compactification* is a compactification which is a Hausdorff space.

The proof of this theorem uses the concept of a diagonal product of a collection of maps and Tikhonov's theorem on the compactness of the topological product of compact spaces (see [E, Chap. III] for more details).

As a rule, every non-compact completely regular T_1-space has many Hausdorff compactifications which are related to one another in subtle ways (see [A3], [E, Chap. III]). However, among all Hausdorff compactifications of a space there are certain canonical ones.

Theorem 15. *Every normal T_1-space X has a Hausdorff compactification βX with the property that the closures in βX of any two disjoint closed subsets of X are disjoint.*

The Hausdorff compactificaton βX of X is determined up to a natural equivalence relation by the property above: it is called the *Stone-Čech compactification* of X.

Theorem 15 is used to prove the following result.

Theorem 16. *Every continuous bounded real-valued function f on a normal T_1-space X extends to a continuous real-valued function on the Stone-Čech compactification βX of X.*

It turns out that every completely regular T_1-space X has a Hausdorff compactification with the property mentioned in Theorem 16. This compactification is also called the *Stone-Čech compactification* of X and is denoted by βX; it is characterized by the extension property up to an obvious equivalence relation on compactifications. The following generalization of theorem 16 is useful.

Theorem 17. *Every continuous map of a completely regular T_1-space X to a compact Hausdorff space (that is, to a compactum) Y extends to a continuous map f of the Stone-Čech compactification βX of X to Y.*

Stone-Čech compactifications have numerous applications in general topology and in its applications. They are the maximal elements in the hierarchy of all Hausdorff compactifications of a given space. Already highly nontrivial is the structure of the Stone-Čech compactification $\beta \mathbb{N}$ of the discrete natural numbers; in spite of being compact, the space $\beta \mathbb{N}$ does not have a single nontrivial convergent sequence. Stone-Čech compactifications play a significant role in the general theory of measure, in the theory of linear topological spaces and in the theory of Banach algebras (see [TF], [Ke], [Wa]). The spaces $\beta \mathbb{N}$ and βX were first constructed by Tikhonov in [T2], but they were first singled out as particular objects for investigation by Stone and Čech (see [S2] and [Č1]).

5.7. Locally Compact Spaces

Definition 6. A topological space X is said to be *locally compact* if every point $x \in X$ has a neighbourhood O_x whose closure $\bar{O}_x$ is compact.

Since topology is connected with limiting processes and the latter are local in nature, the solution of many problems depends only on the local structure of a topological space. In such cases, local compactness is a successful substitute for compactness. Moreover, the class of locally compact spaces is considerably larger than the class of all compact spaces: it includes such classical objects as the real line $\mathbb{R}$, the euclidean spaces $\mathbb{R}^n$ and any open subsets of them. One works constantly with locally compact spaces in the theory of functions of a complex variable and in differential topology where a central role is played by manifolds which are locally compact spaces.

Locally compact Hausdorff spaces can fail to be normal. However, one does have the following very useful result.

Proposition 7. *Every locally compact Hausdorff space is completely regular.*

Proposition 8. *A space X is locally compact and Hausdorff if and only if it can be represented as an open subspace of a compact Hausdorff space.*

Proposition 9. *Every locally compact Hausdorff space is an open subset of any Hausdorff compactification of it.*

One often encounters the following construction in classical mathematics. Adjoining a single point to a line gives a circle (the projective line), adjoining one point to a plane gives a sphere (the *Riemann sphere*), and so on. This construction acuqires a particular simplicity and generality in point set topology. Namely, Aleksandrov observed that the following holds.

Theorem 18. *Every locally compact Hausdorff space X can be obtained by removing a single point from a compact Hausdorff space which is uniquely determined by X.*

5.8. Compactness-Type Conditions on Spaces and Maps. One of the most important conditions of this sort is local compactness which was considered in the preceding section.

The following definition singles out the class of topological spaces whose topologies can be uniquely reconstructed from their compact subspaces.

Definition 7. A Hausdorff space X is called a *k-space* if every set P with the following property is closed in X: the intersection of P with every compact subspace F of X is closed in F.

All locally compact Hausdorff spaces are k-spaces, as are all metrizable spaces and all sequential Hausdorff spaces. Thus, the class of k-spaces is very broad. Questions pertaining to passage to limits often turn out to reduce in k-spaces to the case of compact spaces, a fact which accounts for the essential role that k-spaces play in general topology and its applications. In particular, the class of k-spaces is important in algebraic topology – this is related to the fact that polytopes corresponding to CW-complexes are k-spaces.

Theorem 19 (Cohen). *A Hausdorff space X is a k-space if and only if it is a quotient of a locally compact Hausdorff space* (see [Co]).

Of especial significance is the stability of the property of being a k-space with respect to quotient maps.

Proposition 10. *The image of a k-space under a quotient map to a Hausdorff space is a k-space.*

Unfortunately, the product of two k-spaces need not be a k-space. However, the product of a k-space and a locally compact Hausdorff space is a k-space (see [Co], [E]).

In investigating the topological properties of function spaces an essential role is played by the notion of a $k_{\mathbb{R}}$-space, which is an extension of the concept of a k-space within the framework of the class of completely regular T_1-spaces.

Definition 8. A completely regular T_1-space X is called a $k_{\mathbb{R}}$-*space* if every real valued function $f: X \to \mathbb{R}$ with the following property is continuous on X: the restriction of f to any compact subspace F of X is continuous.

Warning. If a completely regular T_1-space is a k-space, then it is a $k_{\mathbb{R}}$-space, but the converse is false.

Example 5. The topological product (see § 7) of an uncountable set of copies of the real line $\mathbb{R}$ (for example, the space $\mathbb{R}^{\aleph_1}$) is a $k_{\mathbb{R}}$-space which is not a k-space.

It is also very natural to impose compactness conditions as a tool in classifying maps. We shall introduce some of the most important definitions that arise in this way.

Definition 9. A map of one topological space to another is said to be *compact* if the preimage of every point is a compact subspace.

The notion of a compact map is fairly primitive: deeper is the notion of a k-map defined below.

Definition 10. A map $f: X \to Y$ between topological spaces is called a k-*map* if the preimage $f^{-1}(\Phi)$ of every compact subspace Φ of Y is a compact subspace of X and the image $f(F)$ of any compact subspace F of X is a compact subspace of Y.

Bourbaki refers to k-maps as *proper maps*.

Proposition 11. *Every continuous map between compact Hausdorff spaces is a k-map.*

Among all maps, k-maps occupy roughly the same position as do compact spaces among all topological spaces. Here are two of the central results involving k-spaces.

Theorem 20 (see [ArP]). *Every k-map between k-spaces is continuous and closed.*

Theorem 21 (see [ArP]). *Every continuous, closed compact map of a Hausdorff space X to a Hausdorff space Y is a k-map.*

Continuous closed compact maps are called *perfect maps*.

In connection with the concept of a compact map, we mention yet another important class of maps defined in terms of compactness, the so-called *k*-covering maps.

Definition 11 (see [Ar3]). A continuous map f of a topological space X to a topological space Y is called a *k-covering map* if, for each compact subspace Φ of Y, there exists a compact subspace F of X such that $f(F) \supset \Phi$ (equivalently, $f(F) = \Phi$).

All continuous *k*-maps are *k*-covering maps but the converse is false. The following assertion is a useful complement to Theorem 20.

Proposition 12. *A k-covering map of a Hausdorff space X to a k-space Y is a quotient map.*

§6. Uryson's Lemma and the Brouwer-Tietze-Uryson Theorem

6.1. Uryson's Lemma. We shall say that a space X *admits a dense subdivision between disjoint sets F_0 and F_1* if there exists a family $\{\Gamma_r\}$ of open subsets of X indexed by numbers r in an everywhere dense subset D of the segment $[0, 1]$ such that

$$F_0 \subset \Gamma_r \subset \bar{\Gamma}_r \subset \Gamma_{r'} \subset X \backslash F_1 \tag{1}$$

for all $r, r' \in D$ such that $r < r'$.

Lemma 1. *If a space X admits a dense subdivision between disjoint subsets F_0 and F_1, then there exists a continuous function $\varphi: X \to [0, 1]$ such that $\varphi(F_0) = 0$ and $\varphi(F_1) = 1$.*

Proof. An arbitrary point $x \in X$ defines a cut (A_x, B_x) of the set D:

$$A_x = \{r: x \notin \Gamma_r\}, \, B_x = \{r: x \in \Gamma_r\}.$$

Consequently, there exists a unique number $\varphi(x) \equiv \sup A_x = \inf B_x$. This defines a function $\varphi: X \to [0, 1]$. It is clear that $\varphi(F_0) = 0$ and $\varphi(F_1) = 1$. We only need to check that φ is continuous. Since the semi-intervals, $[0, a)$ and $(b, 1]$ are a subbase of the segment $[0, 1]$, it suffices to show that the sets $\varphi^{-1}[0, a)$ and $\varphi^{-1}(b, 1]$ are open. This follows from the identities

$$\varphi^{-1}[0, a) = \bigcup \{\Gamma_r: r < a\}$$

and

$$\varphi^{-1}(b, 1] = \bigcup \{X \backslash \bar{\Gamma}_r: r > b\}$$

which are easily established using condition (1).

Uryson's Lemma. *For any two disjoint closed sets F_0 and F_1 in a normal space X, there exists a continuous function $\varphi: X \to [0, 1]$ such that $\varphi(F_0) = 0$ and $\varphi(F_1) = 1$.*

Proof. By the preceeding lemma, it suffices to prove that there exists a dense subdivision between F_0 and F_1. Take D to be the set of all binary-rational numbers in the segment $[0, 1]$. Set $\Gamma_1 = X \backslash F_1$. Since X is normal, there exists a neighbourhood Γ_0 of F_0 such that $\bar{\Gamma}_0 \subset \Gamma_1$. Then there exists a neighbourhood $\Gamma_{1/2}$ of Γ_0 such that $\bar{\Gamma}_{1/2} \subset \Gamma_1$ and so on.

6.2. The Brouwer-Tietze-Uryson Theorem for Continuous Functions. *Let F be a closed subset of a normal space X and $\varphi: F \to \mathbb{R}$ a continuous function. Then there exists a continuous function $\psi: X \to \mathbb{R}$ such that $\psi|_F = \varphi$. If, in addition, $\varphi(F) \subset [a, b]$, then $\psi(X) \subset [a, b]$.*

One first proves the theorem in the case that the function φ is bounded by applying Uryson's lemma and the theorem that a uniformly convergent sequence of continuous functions on a topological space converges to a continuous function. The case in which φ is unbounded reduces to the case of a function $\varphi: F \to (-1, 1)$. It extends to a function $\psi_1: X \to [-1, 1]$. We set $F_0 = \psi_1^{-1}(-1) \cup \psi_1^{-1}(1)$. By Uryson's lemma, there exists a function $\psi_0: X \to [0, 1]$ such that $\psi_0(F_0) = 0$ and $\psi_0(F) = 1$. Then the function $\psi = \psi_0 \circ \psi_1$ is the desired extension of the function φ.

Remark. Uryson's lemma and the Brouwer-Tietze-Uryson theorem (both the unbounded and bounded parts) actually *characterize* the set of normal spaces.

Corollary 1. *Every continuous map $\varphi: F \to B^n$ of a closed subset F of a normal space X to the closed ball B^n extends to the whole space.*

To prove this, it is necessary to replace the ball B^n by the cube I^n and use the Brouwer-Tietze-Uryson theorem to extend each coordinate function over X.

Corollary 2. *Every continuous map $\varphi: F \to S^n$ of a closed subset F of a normal space X to the sphere S^n extends to a neighbourhood O_F.*

Proof. According to Corollary 1, we can extend φ to a map $\psi_1: X \to B^{n+1}$. Let O be the center of the ball B^{n+1}. Set $O_F = X \backslash \psi_1^{-1}O$ and $\psi = r \circ \psi_1$ where $r: B^{n+1} \backslash \{0\} \to S^n$ is the radial retraction.

§7. Products of Spaces and Maps

7.1. The Topological Product of Spaces. Let $X = \prod \{X_\alpha: \alpha \in A\}$ be *the cartesian product of a collection of sets*; that is, the set of all maps $x: A \to \prod \{X_\alpha: \alpha \in A\}$ where $x(\alpha) \in X_\alpha$.

If $B \subset A$, then there is a natural projection $p_B: X \to \prod \{X_\alpha: \alpha \in B\}$ which associates to each point x of the product (that is, to each mapping $x: A \to \bigcup \{X_\alpha: \alpha \in A\}$) the restriction to the set B. We shall often denote this restriction by p_B^A. If B consists of a single element α, then p_B will be denoted p_α.

If $x \in \prod \{X_\alpha: \alpha \in A\}$, we will call $x(\alpha)$ the αth coordinate of the point x and sometimes denote it by x_α.

Now, suppose that the components X_α of the product $X = \prod \{X_\alpha : \alpha \in A\}$ are topological spaces and consider the smallest topology on X which makes all the maps $p_\alpha : X \to X_\alpha$ continuous (that is, the weak topology generated by the family of maps p_α). The set X with this topology is called the *topological* or *Tikhonov product* or, more simply, the *product of the spaces X_α*.

A subbase of X consists of all sets $p_\alpha^{-1} U$ where U is open in X_α (see §4). Consequently, a base of X consists of all possible finite intersections $p_{\alpha_1}^{-1} U_1 \cap \cdots \cap p_{\alpha_s}^{-1} U_s$.

Proposition 1. *Let X be the topological product of the spaces X_α, $\alpha \in A$ and suppose that $f : Y \to X$ is a map for which all the compositions $p_\alpha \circ f : Y \to X_\alpha$ are continuous. Then f is continuous.*

Proposition 2. *The Tikhonov poduct coincides with the categorical product in the category **Top** of all topological spaces and continuous maps. That is, it has the following property: for every collection $\{q_\alpha : Y \to X_\alpha\}$ of continuous maps, there exists a unique continuous map $h : Y \to X = \prod \{X_\alpha : \alpha \in A\}$ such that $p_\alpha \circ h = q_\alpha$.*

Proposition 3. *The product of the subspaces $Y_\alpha \subset X_\alpha$, $\alpha \in A$, is naturally homeomorphic to the subspace $\bigcap \{p_\alpha^{-1} Y_\alpha : \alpha \in A\}$ of the product $\prod \{X_\alpha : \alpha \in A\}$.*

If all the components are homeomorphic to the same space X, then the product $\prod \{X_\alpha : \alpha \in A\}$ will be denoted by X^A and called a (the A-fold) power of the space X.

Proposition 4. *The product of any collection of T_i-spaces, $i = 0, 1, 2, 3, \rho$ is a T_i-space.*

7.2. Products of Maps. Let $f_\alpha : X_\alpha \to Y_\alpha$, $\alpha \in A$, be maps. Then the map $f : \prod X_\alpha \to \prod Y_\alpha$ defined by the equations $f(x)(\alpha) = f_\alpha(x)$ is called the *product of the maps f_α* and is denoted by $\prod \{f_\alpha : \alpha \in A\}$.

If $f_\alpha : X \to Y_\alpha$, $\alpha \in A$, is a collection of maps, the map $f : X \to \prod Y_\alpha$ defined by the rule $f(x)(\alpha) = f_\alpha(x)$ is called the *diagonal product of the maps f_α* and is denoted by $\Delta \{f_\alpha : \alpha \in A\}$.

Both the product and diagonal product of continuous maps are continuous by Proposition 1.

The *interior product of a collection of maps $f_\alpha : X \to Y_\alpha$, $\alpha \in A$,* is the epimorphism $\bigotimes \{f_\alpha : \alpha \in A\}$ and is equivalent to the diagonal product $\Delta \{f_\alpha : \alpha \in A\}$; that is, $\bigotimes f_\alpha$ coincides with the diagonal product Δf_α considered as a map from the space X to its image.

Thus, if $f_1 = f_2 = id_x$, then $f_1 \Delta f_2$ is the diagonal inclusion of the space X into its square, and $f_1 \otimes f_2$ is a homeomorphism of X to the diagonal $\Delta(X \times X)$.

7.3. Fiber and Fan Products of Mappings and Spaces. Let $f_\alpha : X_\alpha \to Y, \alpha \in A$, be a collection of continuous maps. Consider the diagonal in the product Y^A; namely, consider the diagonal

$$\Delta(Y^A) = \{y = \{y_\alpha\} \in Y^A : y_\alpha = y_\beta \text{ for } \alpha, \beta \in A\}.$$

Let X denote the preimage of the diagonal $\Delta(Y^A)$ under the map $\prod\{f_\alpha: \alpha \in A\}$ and let f_A be the restriction of the map to X. Let δ_Y^A denote the natural homeomorphism of the diagonal $\Delta(Y^A)$ onto Y (the map δ_Y^A associates to a point $\{y_\alpha\} \in \Delta(Y^A)$ any of its coordinates). We let f denote the composition $\delta^A \circ f_A$.

We call X the *fan product of the spaces* X_α over Y with respect to the collection of maps $\{f_\alpha: \alpha \in A\}$ and denote it by $\prod\{X_\alpha, f_\alpha: \alpha \in A\}$. If Y is a single point, then the fan product coincides with the usual product. The map f is called the *fiber product of the collection of maps* $\{f_\alpha: \alpha \in A\}$ and denoted by $\prod\{f_\alpha: X_\alpha \to Y, \alpha \in A\}$.

The restriction $\pi_\beta: X \to X_\beta$ of the projections $p_\beta: \prod_\alpha X_\alpha \to X_\beta$ will be call the *projections of the fan product* X to the components X_β. It is easy to see that $f = f_\alpha \circ \pi_\alpha$ for any $\alpha \in A$. Therefore, the preimage $f^{-1}y$ of any $y \in Y$ coincides with the product $\prod\{f_\alpha^{-1}y: \alpha \in A\}$ and, for any point $x_\alpha \in X_\alpha$, the preimage $\pi_\alpha^{-1}x_\alpha$ coincides with the product $\{x_\alpha\} \times \prod_{\beta \neq \alpha} f_\beta^{-1}f_\alpha x_\alpha$. Thus, the fibers of f are the products of the fibers of the maps f_α.

The fan product of the maps $f_\alpha: X_\alpha \to Y$, $\alpha \in A$, refers to the fiber product f of the maps together with the collection of projections $\pi_\alpha: X \to X_\alpha$. Thus, the fan product of two maps f_1, f_2 is the pullback diagram or Cartesian square

$$
\begin{array}{ccc}
X & \xrightarrow{\;\pi_2\;} & X_2 \\
\pi_1 \downarrow & & \downarrow f_2 \\
X_1 & \xrightarrow{\;f_1\;} & Y
\end{array}
$$

7.4. The Category Top_Y. The objects of this category are continuous maps to a fixed space Y and are sometimes called spaces over the base Y. A morphism between two objects $f_1: X_1 \to Y$ and $f_2: X_2 \to Y$ is a map $g: X_1 \to X_2$ such that $f_2 \circ g = f_1$.

Theorem. *The fan product of the maps* $f_\alpha: X_\alpha \to Y$ *is a product in the category* Top_Y.

Thus, while the nonemptiness of the product of nonempty objects is equivalent to the axiom of choice in the category *Top*, a product of even just two nonempty objects can be empty in the cateogry Top_Y. An example is the pair of maps $f_i: X_i \to Y$, $i = 0, 1$, where $Y = D = \{0, 1\}$ is the discrete (or simple) two point set, $X_i = \{i\}$, and f_i is the natural inclusion.

7.5. Tikhonov's Lemma. *Let* $f_\alpha: X \to Y_\alpha$, $\alpha \in A$, *be a collection of continuous maps which separate points and closed sets of a T_1-space X (that is, for each $x \in X$ and each closed set $F \subset X$ not containing x, there exists an $\alpha \in A$ such that $f_\alpha(x) \notin \overline{f_\alpha(F)}$). Then the diagonal product* $f = \Delta f_\alpha: X \to \prod Y_\alpha$ *is an inclusion.*

Proof. The map $f: X \to f(X)$ is clearly one to one and continuous. We need only prove that it is open. Let Ox be an arbitrary neighbourhood of $x \in X$. By hypothesis, there exists a map $f_\alpha: X \to Y_\alpha$ such that $f_\alpha(X) \notin \overline{f_\alpha(X \backslash Ox)}$. We set $U = Y_\alpha \backslash \overline{f_\alpha(X \backslash Ox)}$. Then the set $V = f(X) \cap p_\alpha^{-1} U$ is a neighbourhood of the point $f(x)$ contained in fOx.

7.6. Examples of Products of Spaces and Maps

Example 1. *The Tikhonov cube* I^τ of weight τ is the product of τ copies of the segment I. It follows from Tikhonov's lemma that every completely regular space can be imbedded in a Tikhonov cube. The converse is also true: every subspace of a Tikhonov cube is completely regular. It can be shown that a Tikhonov cube is compact.

Example 2. *The Hilbert cube* Q is the subspace of the Hilbert space $H^\omega = l_2$ consisiting of all points whose i^{th} component lies in the segment $[0, 1/2^i]$. The Hilbert cube is homeomorphic to the Tikhonov cube I^ω of countable weight. Every subspace of it is metrizable and separable. Conversely, every separable mertizable space imbeds in the Hilbert cube. The latter easily follows from Tikhonov's lemma and the fact that a separable metric space has a countable base.

Example 3. *The Cantor discontinuum (or cube)* D^τ of weight τ is the product of τ copies of the discrete two point space $\{0, 1\}$. It follows immediately from Tikhonov's lemma that every inductively zero-dimensional space X of weight τ imbeds in D^τ (Tikhonov's lemma is applied to the set of characteristic functions of a base of open-closed sets of X of cardinality τ). It is clear that the converse is also true: every subset of the Cantor discontinuum is inductively zero-dimensional. It is also clear that the Cantor cube is homeomorphic to a closed subset of I^τ.

Example 4. The Cantor discontinuum D^ω of countable weight is homeomorphic to the Cantor perfect set (see article II of this volume; Chap. 1, § 3).

Example 5. Let $\mathbb{R}^\tau$ be the product of τ copies of the real line. The space $\mathbb{R}^\tau$ has a natural linear structure and plays a large role in both general topology and functional analysis. For finite $\tau = n$, one obtains n-dimensional Euclidean space. The closed subspaces of $\mathbb{R}^\tau$ exhaust the class of real compact spaces of weight less than or equal to τ. For this reason, real compact spaces are sometimes called $\mathbb{R}$-compacta.

Example 6. Let $\mathbb{N}^\tau$ be the product of τ copies of the set $\mathbb{N}$ of nonnegative real numbers. This space is naturally homeomorphic to a closed subspace of $\mathbb{R}^\tau$ and, consequently, is $\mathbb{R}$-compact. The set $\mathbb{N}$ raised to a countable power can be given a Baire metric and, consequently, is homeomorphic to the Baire space of countable weight (see § 2). It can also be shown that $\mathbb{N}^\omega$ is homeomorphic to the space of all irrational numbers of the real line.

Example 7. If r is a binary rational number belonging to the set R of all binary rational numbers in the interval $(0, 1)$, we let X_r denote the subspace of the real line $\mathbb{R}$ consisting of the two segments $[0, r]$ and $[1 + r, 2]$. Set $Y = [0, 1]$ and let $f_r: X_r \to Y$ denote the map which fixes the segment $[0, r]$ and translates the segment $[1 + r, 2]$ by one unit to the left. Then the fan product X of the spaces X_r over Y with respect to the family of maps $\{f_r: r \in \mathbb{R}\}$ is homeomorphic to the Cantor perfect set and the fiber product $f: X \to Y$ of the collection $\{f_r: r \in \mathbb{R}\}$ is homeomorphic to the natural projection of the Cantor perfect set onto the segment formed by gluing together the endpoints of adjacent intervals (see article II of this book, Chap. 1, § 3).

7.7. The Parallels Lemma. Let the diagram

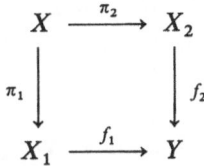

be the fan product of the maps f_1 and f_2.

The First "Parallel" Lemma. *If $x \in X_1$ and $y = f_1 x$, the projection π_2 homeomorphically maps the set $\pi_1^{-1} x = x \times f_2^{-1} y$ to the set $f_2^{-1} y$.*

Thus, the fibers of π_1 are homeomorphic to the fibers of the map f_2. Hence, if the fibers of f_2 possess some common topological property (for example, if all are bicompact, or zero-dimensional, or finite), then the fibers of the "parallel" map π_1 possess the same property.

The Second "Parallel" Lemma. 1) *If f_2 is an open map, then π_1 is also open.* 2) *If the map f_2 is closed and bicompact, then the map π_1 is also closed and bicompact.*

§ 8. Connectedness of Topological Spaces

8.1. Connectedness. A topological space X is said to be *connected* if its only open-closed subsets are $\emptyset$ and X; otherwise, it is said to be *disconnected*. Every disconnected space X can be represented as a disjoint sum $X = A \cup B$ of nonempty open-closed sets A and B. A subset Y of a topological space X is called *connected*, if the space Y is connected in the induced topology.

A *chain of sets* is a finite sequence $M_0, M_1, \ldots, M_s$ of sets with the property that $M_i \cap M_{i+1} \neq \emptyset$ for each i. We shall quote a number of simple facts about connected sets.

Proposition 1. *If $\mathcal{M}$ is a collection of connected subsets any two of which are joined by a chain, then the union $\bigcup \mathcal{M}$ is connected.*

Proposition 2. *Any segment of the number line is connected.*

Proposition 3. *Every convex subset of a real linear topological space is connected.*

Proposition 4. *If Y is a connected set in a space X and if Z is any set satisfying the condition $Y \subset Z \subset \bar{Y}$, then Z is connected. In particular, the closure of a connected set is connected.*

Proposition 5. *A continuous image of a connected set is connected.*

Proposition 6. *A product of connected spaces is connected.*

A nonempty connected subset Y of a space X is said to be a *maximal connected set* or a (*connected*) *component* of X if every connected subset $Z \subset X$ containing Y coincides with Y. Two distinct components of a space X are disjoint by Proposition 1. By Proposition 1 again, each point is contained in a connected component and, by Proposition 4, the connected conmponents are closed. Thus, each space is a disjoint sum of its connected components.

8.2. Local Connectedness. A space X is said to be *locally connected* if its connected open subsets are a base of X. We get an equivalent definition by requiring that each point of X possess a base of connected (not necessarily open) neighbourhoods.

Proposition 7. *A space X is locally connected if and only if the components of its open sets are open.*

In contrast to connectedness, local connectedness is not preserved under continuous maps or by taking products. However, a product of finitely many locally connected spaces is locally connected, It is also evident that the open image of a locally connected space is locally connected. In addition, we have the following.

Proposition 8. *A closed image of a locally connected space is locally connected.*

A connected metrizable compactum is called a *continuum*. Propositions 5 and 7 imply the following.

Proposition 9. *The property of being a continuum or being a locally connected continuum is preserved by continuous maps to Hausdorff spaces.*

8.3. Local Path Connectedness. A space X is said to be 1) *path connected* if any two points of X can be joined by a path and 2) *locally path connected* if it has a basis of open path connected sets. An equivalent definition of local path connectedness is obtained by requiring that, for each point $x \in X$ and every neighbourhood Ox of x, there exist a smaller neighbourhood $O_1 x$ any two points of which can be joined by a path in Ox. One also gets equivalent definitions of path and local path connectedness by replacing arcs by continuous (Hausdorff) images of segments.

It is evident that every (local) path connected space is (locally) connected. Connectedness does not imply path connectedness, even for continua. Concerning local connectedness, we have the following.

Proposition 10. *Every locally connected complete metric space is locally path connected.*

It is also clear that every connected, local path connected space is path connected.

§9. Inverse Systems, Spectra of Topological Spaces and their Limits

9.1. Definition of Inverse Systems and Spectra. A category $C = (\mathcal{O}, \mathcal{M})$ is said to be *small* if the collection $\mathcal{O}$ of objects and the collection $[X, Y]$ of morphisms from X to Y are sets for any objects $X, Y \in \mathcal{O}$. A small category $C = (\mathcal{O}, \mathcal{M})$ is called an *inverse* or *projective system* if, for any two objects X, Y,

a) the set $[X, Y]$ contains no more than one element.

If, in addition, the category C consists of sets and maps (respectively, topological spaces and continuous maps) then it is called an *inverse* or *projective system of sets* (respectively, *topological spaces*). If a category C is an inverse system then its objects are called *elements* and its morphisms are called *projections* of the inverse system.

Remark 1. The definition of a direct or inductive system of sets (topological spaces) is no different from the definition of an inverse system. The difference only begins when we define limits of these systems.

The collection of objects of any category admits a natural *partial pre-order*; that is, a reflexive and transitive binary relation $\leqslant$. Namely,

$$Y \leqslant X \quad \text{if and only if} \quad [X, Y] \neq \varnothing. \tag{1}$$

An inverse system $C = (\mathcal{O}, \mathcal{M})$ is called an *inverse spectrum* if

b) the partially pre-ordered set $\mathcal{O}$ is *directed* (*above*); that is, for any two elements $X, Y \in \mathcal{O}$ there exists an element $Z \in \mathcal{O}$ such that $X \leqslant Z$ and $Y \leqslant Z$.

If $X, Y \in \mathcal{O}$ and $Y \leqslant X$, then condition a) mandates that there exist a unique projection from X to Y which we denote by π_Y^X. We sometimes use another Greek or Latin letter in place of π. Since an inverse system is a category, if $Z \leqslant Y \leqslant X$ we have

$$\pi_Z^X = \pi_Z^Y \circ \pi_Y^X. \tag{2}$$

Since each inverse system $C = (\mathcal{O}, \mathcal{T})$ is a category, the natural map π_X^X must coincide with id_X for every object $X \in \mathcal{O}$. Therefore, if $X \leqslant Y$ and $Y \leqslant X$, then (2) implies that

$$\pi_Y^X \circ \pi_X^Y = id_Y, \qquad \pi_X^Y \circ \pi_Y^X = id_X.$$

In this case, the maps π_Y^X and π_X^Y are isomorphisms in the category C. Therefore, the equivalence relation (3) on the pre-ordered set $\mathcal{O}$ given by

$$X \sim Y \quad \Leftrightarrow \quad X \leqslant Y \text{ and } Y \leqslant X \tag{3}$$

is finer than the relation of homeomorphism; that is, any two equivalent spaces are homemorphic. In this connection, if $X_1 \sim X_2$, $Y_1 \sim Y_2$ and if $\pi_{Y_i}^{X_i}$, $i = 1, 2$, are projections of the inverse system C, then it follows from the equalities

$$\pi_{Y_2}^{Y_1} \circ \pi_{Y_1}^{X_1} = \pi_{Y_2}^{X_1} = \pi_{Y_2}^{X_2} \circ \pi_{X_2}^{X_1}$$

that the mappings $\pi_{Y_1}^{X_1}$ and $\pi_{Y_2}^{X_2}$ are homeomorphisms. Consequently, the equivalence relation on the set of objects $\mathcal{O}$ extends to the set of morphisms $\mathcal{M}$ (identifying an object X with the morphism $\pi_X^X = id_X$ defines a natural inclusion $\mathcal{O} \subset \mathcal{M}$). The quotient $C' = (\mathcal{O}', \mathcal{M}')$ of an inverse system C is clearly an inverse system. Informally, the systems C and C' can be identified without losing any essential information. Each element of the system C' corresponds to a collection of homeomorphisms and, hence, elements of the system C which are not distinct. The system C' is more convenient because it is "smaller" than C and the set $\mathcal{O}'$ of its objects is partially ordered. Therefore, in what follows we shall suppose that relation (1) is a partial ordering on the set of objects of an inverse system.

It often turns out to be more convenient to denote the elements of an inverse system or spectrum by a single letter, for example X, indexed by elements α of a partially pre-ordered set A. In this case, the projections from X_α to X_β are denoted by π_β^α and the spectrum itself by

$$S = \{X_\alpha, \pi_\beta^\alpha, A\}.$$

In what follows we shall refer to inverse spectra simply as *spectra*.

9.2. Limits of Inverse Systems. Suppose that an inverse system $S = \{X_\alpha, \pi_\beta^\alpha, A\}$ is a subcategory of a category K. Then an object $X \in K$ and a family of morphisms $\pi_\alpha: X \to X_\alpha$, $\alpha \in A$, is said to be a *limit of the system S* in the category K if

$$\pi_\beta^\alpha \circ \pi_\alpha = \pi_\beta \qquad \text{for all } \alpha, \beta \in A, \alpha \leqslant \beta \qquad (4)$$

and if, for any other object $Y \in K$ and family of morphisms $f_\alpha: Y \to X_\alpha$ with the property

$$\pi_\beta^\alpha \circ f_\alpha = f_\beta, \qquad (5)$$

there exists a unique morphism $f: Y \to X$ such that $f_\alpha = \pi_\alpha \circ f$ for all $\alpha \in A$.

The limit is denoted by $\lim KS$ or simply by $\lim S$. The object X is also called the *limit* of the inverse system S and the morphisms π_α the *limiting projections* of the inverse system S.

Remark 2. An example of an inverse system is a family of topological spaces not connected by any maps. The limit of this inverse system is the product of the spaces.

An inverse system has at most one limit and, as is the case with products, the limit need not exist in every category.

In what follows, the limit of an inverse system of topological spaces will mean the limit in the category *Top*. In this category, the limit always exists.

For an inverse system $S = \{X_\alpha, \pi_\beta^\alpha, A\}$ the limit is a subset of the product $X = \prod\{X_\alpha : \alpha \in A\}$ and consists of all points $x \in X$ satisfying the condition

$$p_\beta(X) = \pi_\beta^\alpha p_\alpha(x) \tag{6}$$

for $\beta \leqslant \alpha$. These points are called *threads of the inverse system*. The limiting projections $\pi_\alpha : \lim S \to X_\alpha$ are the restrictions of the projections $p_\alpha : X \to X_\alpha$ to the limit $\lim S$.

The limit of an inverse system of T_i-spaces, $i = 0, 1, 2, 3, \rho$, is a T_i-space. The limit of an inverse system of (completely) regular spaces is (completely) regular. This implies that the enumerated properties are preserved both under taking products and passing to subspaces.

The limit of an inverse system $S = \{X_\alpha, \pi_\beta^\alpha, A\}$ of Hausdorff spaces X_α is closed in the product $X = \prod\{X_\alpha : \alpha \in A\}$. This follows because we have

$$\lim S = \bigcap\{X^{\beta\alpha} : \alpha, \beta \in A, \beta \leqslant \alpha) \tag{7}$$

by (6) and each set $X^{\beta\alpha} = \{X \in X : p_\beta(X) = \pi_\beta^\alpha p_\alpha(X)\}$ is closed as a set on which two maps into the Hausdorff space X_β coincide. This, together with Tikhonov's theorem on the compactness of products, implies the following.

Proposition 1. *The limit of an inverse system of compacta is a compactum.*

But in contrast to the original Tikhonov theorem, the limit of an inverse system of compact spaces might fail to be compact.

Kurosh's Theorem. *The limit of an inverse spectrum $S = \{X_\alpha, \pi_\beta^\alpha, A\}$ of non-empty compacta is a nonempty compactum.*

Proof. Set $\prod_\alpha = \bigcap_{\beta < \alpha} X^{\beta\alpha}$. The set $\prod_\alpha$ is closed in the product $X = \prod\{X_{\alpha'} : \alpha' \in A\}$. Furthermore, $\prod_\alpha$ is nonempty because it coincides with the product $(\prod_{\beta \leqslant \alpha} \pi_\beta^\alpha(X_\alpha)) \times \prod_{\beta \not\leqslant \alpha} X_\beta$. Thus, $\{\prod_\alpha\}$ is a centered system of nonempty closed subsets of the compactum X. By (7) the nonempty compactum $\bigcap_\alpha \prod_\alpha$ coincides with $\lim S$ and this concludes the proof.

The directedness of the preordered set A is essential in this theorem. The theorem does not extend to inverse systems. For example, the limit will be empty for any inverse system $S = \{X_\alpha, \pi_\beta^\alpha, A\}$ for which there exist $\alpha_1, \alpha_2, \beta \in A$ with the properties:

1) $\beta \leqslant \alpha_1, \alpha_2$;
2) $\pi_\beta^{\alpha_1}(X_{\alpha_1}) \cap \pi_\beta^{\alpha_2}(X_{\alpha_2}) = \varnothing$;
3) α_1, α_2 do not have a common successor.

9.3. Morphisms of Inverse Systems and Spectra and their Limits. A *morphism* of an inverse system $S = \{X_\alpha, \pi_{\alpha'}^\alpha, A\}$ to an inverse system $T = \{Y_\beta, \rho_{\beta'}^\beta, B\}$ is a pair (μ, Φ) where $\mu : B \to A$ is a map and $\Phi = \{f_\beta : X_{\mu(\beta)} \to Y_\beta, \beta \in B\}$ is a set of continuous maps with the property that as soon as $\beta \geqslant \beta'$, there exists $\alpha \geqslant \mu(\beta)$, $\mu(\beta')$ such that

$$f_{\beta'} \circ \pi_{\mu(\beta')}^\alpha = \rho_{\beta'}^\beta \circ f_{\beta'} \circ \pi_{\mu(\beta)}^\alpha. \tag{8}$$

We sometimes refer to the family Φ of maps as a morphism of S to T when the map $\mu\colon B \to A$ is understood.

Proposition 2. *Let* $S_1 = \{X_\alpha, {}^1\pi_{\alpha'}^\alpha, A\}$, $S_2 = \{Y_\beta, {}^2\pi_{\beta'}^\beta, B\}$, *and* $S_3 = \{Z_\gamma, {}^3\pi_{\gamma'}^\gamma, C\}$ *be three inverse systems and let* $(\mu_1, \Phi_1)\colon S_1 \to S_2$ *and* $(\mu_2, \Phi_2)\colon S_2 \to S_3$ *be morphisms, where* $\Phi_1 = \{{}^1f_\beta\colon X_{\mu(\beta)} \to Y_\beta, \beta \in B\}$ *and* $\Phi_2 = \{{}^2f_\gamma\colon X_{\mu(\gamma)} \to Z_\gamma, \gamma \in C\}$. *Then the pair* $(\mu_1 \circ \mu_2, \Phi_2 \circ \Phi_1)$ *where* $\Phi_2 \circ \Phi_1 = \{{}^2f_\gamma \circ {}^1f_{\mu_2(\gamma)}\colon X_{\mu_1\mu_2(\gamma)} \to Z_\gamma,$ $\gamma \in C\}$ *is a morphism of* S_1 *to* S_3.

It is easy to check this definition of compositon of morphisms makes the collection of inverse systems and their morphisms into a category which we will denote by *Inv*. The category *Sp* of inverse spectra is a full subcategory of *Inv*.

Theorem 1. *The operation* lim *of taking the limit of an inverse system extends to a covariant functor* lim*: Inv → Top.*

Proof. Let (μ, Φ) be a morphism of the inverse system $S = \{X_\alpha, \pi_{\alpha'}^\alpha, A\}$ to the inverse system $T = \{Y_\beta, \rho_{\beta'}^\beta, B\}$ and let $x = (x_\alpha)$ be a thread of S. For each $\beta \in B$, we set $y_\beta = f_\beta(x_{\mu(\beta)})$ and prove that $y = (y_\beta)$ is a thread of the inverse system T. It is necessary to verify that condition (6) of the definition of thread is satisfied or, what is the same,

$$y_{\beta'} = \rho_{\beta'}^\beta(y_\beta) \tag{9}$$

for $\beta' \leqslant \beta$. By the definition of a morphism of inverse systems, there exists an $\alpha \geqslant \mu(\beta), \mu(\beta')$ such that (8) holds. Substituting the element x_α of the thread x into this equality gives (9). Thus the map $f\colon \lim S \to \lim T$ is well-defined. It is easy to see that

$$f = \prod \{\Delta\{f_\beta\colon \mu(\beta) = \alpha\}\colon \alpha \in \mu(B) \circ p_{\mu(B)}^A |\lim S\},$$

where $p_{\mu(B)}^A\colon \prod\{X_\alpha\colon \alpha \in A\} \to \prod\{X_\alpha\colon \alpha \in \mu(B)\}$ is the projection of the product. It follows that the map $f = \lim(\mu, \Phi)$ is continuous. A direct verification establishes that the operation lim*: Inv → Top* defined above is functorial. This completes the proof of Theorem 1.

We note that the definition of the map $f = \lim(\mu, \Phi)$ implies that

$$\rho_\beta \circ f = f_\beta \circ \pi_{\mu(\beta)}. \tag{10}$$

This equality will be important in what follows.

Remark 3. It is clear that the map $f\colon \lim S \to \lim T$ is uniquely determined by requiring that condition (10) be satisfied for each $\beta \in B$. Thus, the limit of a morphism of inverse systems is uniquely determined.

We now cite some examples of morphisms of inverse systems.

Example 1. Let the inverse system T be a subsystem of a system S, the map $\mu\colon B \to A$ a topological inclusion and each map $f_\beta\colon X_\beta \to X_\beta$ the identity homeomorphism. This example shows that the limit of an inverse system projects in a natural manner to the limit of any subsystem.

Example 2. Take S to be an inverse spectrum which is a cofinal subspectrum of a spectrum T: that is, $S \subset T$ and the set A is cofinal in B. The map $\mu: B \rightarrow A$ extends the identity map on A. For $\beta \in B \backslash A$, we take $\mu(\beta) \in A$ to be any element $\geqslant \beta$. Such an element exists because A is cofinal in B. We take $f_\beta: X_{\mu(\beta)} \rightarrow X_\beta$ to be the projection of the spectrum $\pi_\beta^{\mu(\beta)}$. Directedness of A ensures that property (8) is satisfied. It is not difficult to verify that the limit $\lim \Phi$ of the morphism $\Phi: S \rightarrow T$ is a homeomorphism. This fact is usually formulated as follows.

Proposition 3. *The limit of an inverse spectrum is homeomorphic to the limit of any cofinal subspectrum.*

Example 3. The morphisms of spectra (inverse systems) which arise most frequently in practice are defined on the same index set A. In this case, the map $\mu: A \rightarrow A$ is taken to be the identity and condition (8) is replaced by the stronger condition

$$f_{\beta'} \circ \pi_{\beta'}^\beta = \rho_{\beta'}^\beta \circ f_\beta \tag{11}$$

The collection of inverse systems (resp., spectra) over a fixed index set A with morphisms defined in this way (the so-called *exact morphisms*) forms a subcategory of the category *Inv* (resp., *Sp*) which we denote by *Inv$_A$* (resp., *Sp$_A$*).

An exact morphism $\Phi: S \rightarrow T$ of inverse systems is called an *isomorphism* if each $f_\alpha \in \Phi$ is a homemorphism. It is clear that a limit of isomorphisms is a homeomorphism.

The following assertion describes still another instance of morphisms of inverse systems.

Proposition 4. *Let $S = \{X_\alpha, \pi_\beta^\alpha, A\}$ be an inverse system, X a space, and $f_\alpha: X \rightarrow X_\alpha$, $\alpha \in A$, a collection of maps satisfying the condition*

$$f_\beta = \pi_\beta^\alpha \circ f_\alpha \tag{12}$$

for all $\alpha, \beta \in A$, $\beta \leqslant \alpha$. Then there exists a natural map $f: X \rightarrow \lim S$ satisfying the condition

$$f_\alpha = \pi_\alpha \circ f,$$

for all $\alpha \in A$ where $\pi_\alpha: \lim S \rightarrow X_\alpha$ is a limiting projection of the inverse system S.

Indeed, the space X, together with its identity map, can be considered as an inverse system over a one element index set. Then the collection $\Phi = \{f_\alpha: \alpha \in A\}$ is a morphism of X to S by (12). Applying Theorem 1 and Remark 3 completes the proof of Proposition 4. The map f is called the *limit of the maps f_α* and is denoted $\lim\{f_\alpha: \alpha \in A\}$.

9.4. Examples of Inverse Systems and their Limits. We already noted above that a product of spaces is the limit of inverse systems whose projections are identity maps of the spaces. There is a close connection between products of spaces and inverse spectra which will be discussed in § 10.

Example 1. A collection F of maps $f_\alpha : X_\alpha \to Y$, $\alpha \in A$, of spaces X_α into a fixed space Y is an inverse system whose limit coincides with the fan product of the spaces X_α over Y with respect to the collection of maps F.

An inverse spectrum whose index set is the set $\mathbb{N}$ of nonnegative integers is called an *inverse sequence*. We cite three examples of inverse sequences $S = \{X_n, \pi_n^{n+1}, \mathbb{N}\}$.

Example 2. Let X_n consist of 2^n isolated points $\{1, \ldots, 2^n\}$, let the projection π_n^{n+1} be the identity on X_n and carry the point $2^n + k$ to k, $1 \leqslant k \leqslant 2^n$. Then $\lim S$ is homeomorphic to the Cantor perfect set.

Example 3. Let X_n be the segment $[0, 1]$ and let all the projections π_n^{n+1} be the same and have the graph sketched in Fig. 1. Then $\lim S$ is a continuum which is imbedded in the plane and which can be described as a intersection of a sequence of compact sets whose first three elements are sketched in Fig. 2. The thickness of the strips tends to zero, and each succeeding strip runs twice around the preceding.

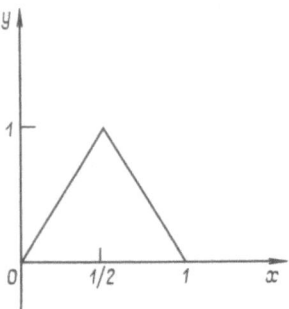

Fig. 1

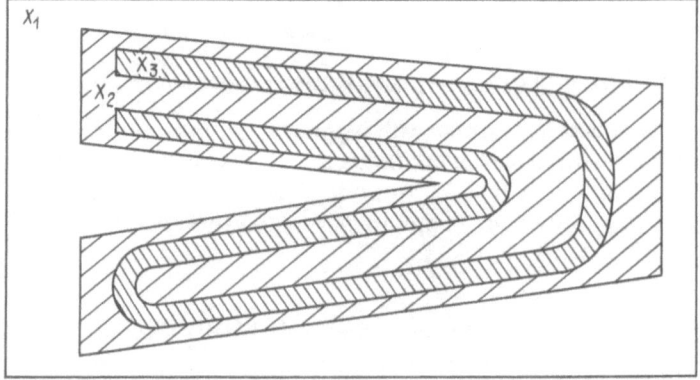

Fig. 2

Example 4. Let X_n be the unit circle in the complex plane, and let each projection π_n^{n+1} be given by squaring. Then $\lim S$ is a continuum which is imbedded in $\mathbb{R}^3$. It is called a *solenoid* and can be represented as an intersection of a decreasing sequence of sets homemorphic to solid tori. Each successive solid torus "runs around" the preceeding one twice.

§ 10. The Relation Between Inverse Spectra and Products

In this chapter, we shall for brevity refer to inverse spectra as *spectra*. An element α of a partially ordered set A is called a *limiting element* if for any two elements, $\beta_1, \beta_2 \in A$, $\beta_i < \alpha$, there exists $\gamma \in A$ such that $\beta_i < \gamma < \alpha$, $i = 1, 2$. If $\alpha \in A$ is a limiting element, the set $A|_\alpha = \{\beta \in A: \beta < \alpha\}$ is directed. If $S = \{X_\alpha, \pi_\beta^\alpha, A\}$ is a spectrum and γ is a limiting element of A, then $S|_\gamma$ will denote the restriction of S to the set $A|_\gamma$. A *spectrum* $S = \{X_\alpha, \pi_\beta^\alpha, A\}$ is said to be *continuous* if, for each limiting element $\gamma \in A$, the limit map $\lim_{\alpha < \gamma} \pi_\alpha^\gamma \colon X \to \lim(S|_\gamma)$, which exists by Proposition 4 of § 8, is a homeomorphism.

Proposition 1. *Every spectrum is isomorphic to a cofinal part of a continuous spectrum.*

Indeed, let $S = \{X_\alpha, \pi_\beta^\alpha, A\}$ be a spectrum. Let $\bar{A}$ denote the set obtained from A by putting in "doubles" $\bar{\alpha}$ of the limiting elements $\alpha \in A$. Put $\alpha < \bar{\alpha}$ and $\beta < \bar{\alpha}$ for all $\beta < \alpha$. The set A is clearly cofinal in $\bar{A}$. The desired continous spectrum $\bar{S}$ is obtained from S by supplementing it with the spaces $X_{\bar{\alpha}} = \lim(S|_\alpha)$, for each $\alpha \in A$, and the corresponding projections.

A collection $\mathcal{D} = \{D\}$ of subsets of a set A is said to be a *direction* in A if
1) $\mathcal{D}$ is a cover of A;
2) the collection $\mathcal{D}$ is *directed by inclusion*; that is, for any $D, D' \in \mathcal{D}$, there exists $D'' \in \mathcal{D}$ which contains D and D'.

Example 1. If the directions in A are considered as covers, then among them there is a finest direction: that is, the direction $\mathcal{D}$ which is a refinement of every other direction $\mathcal{D}'$. The direction $\mathcal{D}$ consists of all finite subsets of A.

Example 2. We obtain another direction if we consider a well-ordering w of $A = \{\alpha: \alpha < \beta\}$ and set $D_\alpha = \{\gamma: \gamma < \alpha\}$. The collection $\mathcal{D} = \{D_\alpha: \alpha < \beta\}$ is also a direction, which is called the (well-ordered) direction associated the well-ordering w on A.

Let $\{X_\alpha: \alpha \in A\}$ be a set of topological spaces and $\mathcal{D}$ a direction in A. For each $D \in \mathcal{D}$, set $X_D = \prod\{X_\alpha: \alpha \in D\}$ and, for $D, D' \in \mathcal{D}$, $D' \subset D$, let $p_{D'}^D \colon X_D \to X_{D'}$ denote the natural projection of the product to the subproduct (see § 7). The collection $\{X_D, p_{D'}^D, \mathcal{D}\}$ is an inverse spectrum called the *spectrum over the direction $\mathcal{D}$* and is denoted $S_\mathcal{D}$.

Theorem 1. *If* $\{X_\alpha : \alpha \in A\}$ *is a collection of topological spaces and* $\mathscr{D}$ *is a direction in the index set* A, *then* $S_\mathscr{D}$ *is a continuous spectrum, its limit coincides with the product* $\{X_\alpha, \alpha \in A\}$, *and the limiting projections of* $S_\mathscr{D}$ *coincide with the projections* $p_\mathscr{D}$ *of the product.*

We shall say that maps $f_i : X_i \to Y_i$, $i = 1, 2$, are *homeomorphic* if there exist homeomorphisms $g : X_1 \to X_2$ and $h : Y_1 \to Y_2$ such that $h \circ f_1 = f_2 \circ g$.

From Theorem 1 we obtain the following.

Corollary 1. *The Tikhonov cube* I^τ *is*
a) *the limit of the spectrum of finite-dimensional cubes* I^n;
b) *the limit of a continuous well-ordered spectrum of Tikhonov cubes* $I^{\tau'}$ *of weight* $\tau' < \tau$.
In addition, if $\tau = \omega_{\alpha+1}$, *the cube* I^τ *is the limit of a continuous spectrum all of whose projections are homemorphic to the projection* $I^{\omega_\alpha} \times I^{\omega_\alpha} \to I^{\omega_\alpha}$.

Corollary 2. *The Cantor discontinuum* D^τ *is:*
a) *the limit of a spectrum of finite spaces;*
b) *the limit of a continuous well-ordered spectrum of Cantor discontinua* $D^{\tau'}$ *of weight* $\tau' < \tau$.
For $\tau = \omega_{\alpha+1}$, *the discontinuum* D^τ *is the limit of a continuous spectrum all projections of which are homeomorphic to the projection* $D^{\omega_\alpha} \times D^{\omega_\alpha} \to D^{\omega_\alpha}$.

From Corollary 1 and Tikhonov's theorem on imbedding completely regular spaces in I^τ, we obtain the following.

Theorem 2. *Every compact set* X *of weight* τ *is:*
a) *the limit of a spectrum of finite dimensional metrizable compacta;*
b) *the limit of a continuous, well-ordered spectrum of compacta of weight less than* τ.

§11. The Spectral Representation Theorem for Mappings

11.1. The Factorization Lemma for Sigma-Complete Spectra. A partially ordered set A is said to be τ-*complete* if every subset C of cardinality less than or equal to τ has a strict upper bound in A. For $\tau = \omega_0$ we will say that τ-complete sets are *sigma-complete*. A spectrum over a τ-complete (respectively, sigma-complete) set is said to be a τ-*complete* (resp., *sigma-complete*) *spectrum*.

Lemma 1. *Let* $S = \{X_\alpha, \pi_\beta^\alpha, A\}$ *be a sigma-complete spectrum of compacta. Then, for each continuous function* $\varphi : \lim S \to \mathbb{R}$, *there exists a continuous function* $\psi : X_\alpha \to \mathbb{R}$, $\alpha \in A$, *such that* $\varphi = \psi \circ \pi_\alpha$.

The proof of this lemma is easily obtained by applying the Stone-Weierstrass theorem. From Lemma 1, we immediately obtain the following.

Factorization Lemma. *Let* $S = \{X_\alpha, \pi_\beta^\alpha, A\}$ *be a* τ-*complete spectrum of compacta and suppose that* Y *is a compactum of weight less than or equal to* τ. *Then* *for every continuous map* $f: \lim S \to Y$ *there exists a continuous map* $g: X_\alpha \to Y$, $\alpha \in A$, *such that* $f = g \circ \pi_\alpha$.

Definition. The *weight of a map* $f: X \to Y$ is the least cardinal number τ such that f imbeds in a projection $p_Y: Y \times Z \to Y$ where $wZ = \tau$. In this case we write $wf = \tau$.

The weight of any map $f: X \to Y$ is well-defined and does not exceed the weight of the space X. Indeed, the map $x \to (x, f(x))$ effects an imbedding of the map f into the projection $p_Y: Y \times X \to Y$.

We say that a map $f: X \to Y$ *factors* in a spectrum $S = \{X_\alpha, f_\beta^\alpha, A\}$ if A has least element 0 and f is homeomorphic to the limiting projection f_0 of the spectrum S.

The proof of the following parameterized version of the factorization lemma is also based on Lemma 1.

The Factorization Lemma for Mappings. *Suppose that* $f: X \to Y$ *is a map which factors in a* τ-*complete spectrum* $S = \{X_\alpha, f_\beta^\alpha, A\}$ *of compacta. Then for each map* $h: X \to Z$ *of* f *into* $g: Z \to Y$ *of weight* $wg \leqslant \tau$ *there exists an index* $\alpha \in A$ *and a map* $k: X_\alpha \to Z$ *such that* $h = k \circ f_\alpha$.

The most interesting application of this lemma occurs for $\tau = \omega_0$; that is, for sigma-complete spectra. This is connected with the fact that each map $f: X \to Y$ of compacta factors in a sigma-complete spectrum of maps of countable weight (such spectra are called *sigma-spectra*). Indeed, if $wf = \tau$, then f imbeds in the projection $p_Y: Y \times I^\tau \to Y$. We take the index set A to be the collection of at most countable subsets $a \subset \tau$. We put $X_a = (id_Y \times p_a)(X)$ and $f_b^a = id_Y \times p_b^a | X_a$.

11.2. The Spectral Theorem for Mappings. Let $S = \{X_\alpha, \pi_\beta^\alpha, A\}$ *be a continuous sigma-complete spectrum,* $T = \{Y_\alpha, \rho_\beta^\alpha, A\}$ *a sigma-spectrum, and* X *and* Y *their limits, respectively. Suppose that the maps* $f_0: X_0 \to Y_0$ *and* $f: X \to Y$ *satisfy the condition*

$$\rho_0 \circ f = f_0 \circ \pi_0.$$

Then the map f *is the limit of an exact morphism between closed cofinal subspectra of* S *and* T.

The proof of the theorem consists in applying the factorization lemma for mappings to check closure and cofinality in A of the set of α for which the mappings $\rho_\alpha \circ f \circ \pi_\alpha^{-1}$ are single-valued.

Corollary. *In the category* **Top$_Y$**, *a homeomorphism of two mappings which factor in sigma-spectra over the same directed set is generated by an isomorphism of closed cofinal subspectra.*

Remarks. 1) All spaces in this section were assumed compact, although the assertions are true under weaker assumptions. 2) Both the theorem and the corollary are also true for τ-spectra. 3) For the one-point space Y the corollary is the well-known *Shchepin spectral theorem for homemorphisms*.

§ 12. Some Concepts and Facts of Uniform Topology

12.1. Definitions of Uniform Spaces and Uniformly Continuous Mappings.
More details on the concepts connected with covers (refinement, star refinement,
local finiteness, and so forth) which are not defined here can be found in [E]. We
use the expressions "the cover u is a refinement of v" and "the cover u refines the
cover v" interchangeably. If a cover u is a refinement (resp., star refinement) of
the cover v, we will write $u > v$ or $v < u$ (resp., $u* > v$ or $v <* u$).

A *uniform structure* or *uniformity* on a set X is a collection $\mathscr{U}$ of covers of X
satisfying the following properties:

U_1) if $u \in \mathscr{U}$ and $u > v$, then $v \in \mathscr{U}$;

U_2) if $u_1, u_2 \in \mathscr{U}$, then there exists a $v \in \mathscr{U}$ such that $v > u_i$, $i = 1, 2$;

U_3) if $u \in \mathscr{U}$, then there exists a cover $v \in \mathscr{U}$ such that $v* > u$.

The pair $(X, \mathscr{U})$ where $\mathscr{U}$ is a uniformity on a set X is called a *uniform space*.
Sometimes, when the uniformity $\mathscr{U}$ on the set X is fixed, the uniform space $(X, \mathscr{U})$
is denoted by X.

A collection $\mathscr{U}_1$ of covers of a space X is called a *basis of a uniformity* $\mathscr{U}$ if
every cover $u \in \mathscr{U}$ has a refinement $u_1 \in \mathscr{U}_1$. A basis of a uniformity satisfies
conditions U_2) and U_3). Conversely, any collection of covers satisfying U_2) and
U_3) is the basis of a (uniquely determined) uniformity.

Let $(X, \mathscr{U})$ be a uniform space. Consider the collection $\mathscr{T}(\mathscr{U})$ of all subsets T
of X with the property that, for every $x \in T$, there exists a cover $u \in \mathscr{U}$ such that
$St_u X \subset T$. It is easy to see that the collection $\mathscr{T}(\mathscr{U})$ is a topology on the set X;
it is called the *topology associated to the uniformity* $\mathscr{U}$ or the *uniform topology*.

If $(X, \mathscr{T})$ is a topological space, then a uniformity $\mathscr{U}$ on X is called *compatible*
with the topology $\mathscr{T}$ if $\mathscr{T} = \mathscr{T}(\mathscr{U})$. In this case, open uniform covers generate a
basis of the uniformity $\mathscr{U}$.

If $\mathscr{U}$ is a uniformity on X and $Y \subset X$, then the collection $\mathscr{U}|Y$ consisting of all
traces of covers $u \in \mathscr{U}$ on the set Y is a uniformity on Y which is called the
uniformity *induced by* $\mathscr{U}$.

A mapping $f: (X, \mathscr{U}) \to (Y, \mathscr{V})$ of uniform spaces is called *uniformly continuous*
if $f^{-1} v \in \mathscr{U}$ for every $v \in V$.

Example 1. A uniform cover of a metric space is any cover u refined by a cover
consisting of all ε-balls for some $\varepsilon > 0$. The family of all uniform covers of a metric
space X is a uniformity called the *metric uniformity* of X.

Example 2. The set of all covers of a set X is a uniformity on X. This uniformity generates the discrete topology on X and has a basis consisting of a single
cover.

Example 3. The set of all finite covers of a set X is a uniformity on X. This
uniformity also generates the discrete topology.

Remark 1. Examples 2 and 3 show that the same topology can be generated
by different uniformities.

Example 4. A cover u of a topological space X is called *normal* if there exists a sequence of open covers u_1, u_2, ... such that $u_1 \, ^* > u$, ... $u_{i+1} \, ^* > u_i$. The collection of all normal covers is a uniformity called the *fine* or *universal uniformity* of the topological space X. If uniformities are compared by inclusion, then the fine uniformity on X is the largest uniformity compatible with the topology on X. A basis of the fine uniformity of a normal topological space X is generated by all open, locally finite covers of X.

12.2. Uniformities and Pseudometrics

Theorem 1 (The subordinate pseudometric lemma). *For each uniform cover u of a uniform space X there exists a metric space (M^u, ρ) and a uniformly continuous map $f_u: X \to M^u$ such that the collection of all preimages of balls of radius 1 is a refinement of u.*

If we set $d_u(x, y) = \rho(f_u(x), f_u(y))$, we obtain a pseudometric (that is, a function satisfying all the metric axioms, except the identity axiom) on X, which is called the *pseudometric subordinate to the uniform cover u.*

Remark 2. It follows from Theorem 1 that the topology of any uniform space is completely regular. Moreover, if a uniform space is separated (that is, if its topology is Hausdorff), then it is Tikhonov.

Define an equivalence relation $\sim_{\mathscr{U}}$ on a uniform space $(X, \mathscr{U})$ as follows:

$$x \sim_{\mathscr{U}} y \quad \Leftrightarrow \quad St_u x \cap St_u y \neq \varnothing \text{ for any } u \in \mathscr{U}.$$

Let $X_{\mathscr{U}}$ denote the quotient set with respect to this equivalence relation and $g = g_{\mathscr{U}}$ the quotient map.

If $f: X \to Y$ is a map and $U \subset X$ a subset, we let $f \# U$ denote the small image of the set U; that is, the set $\{y \in Y: f^{-1} y \subset U\}$. For the collection u of subsets of a set X we set $f \# u = \{f \# U: U \in u\}$.

The Lemma on the Associated Separated Uniformity. *If $(X, \mathscr{U})$ is any uniform space, the pair $(X_{\mathscr{U}}, g \# \mathscr{U})$ where $g \# \mathscr{U} = \{g \# u: u \in \mathscr{U}\}$ is a uniform space with a Hausdorff topology.*

The uniformity $g \# \mathscr{U}$ is called the *separated uniformity* associated to the uniformity $\mathscr{U}$. It is characterized by the following property: for each uniformly continuous map f of X to a separated uniform space Y, there exists a unique uniformly continuous map $h: X_{\mathscr{U}} \to Y$ for which $f = h \circ g$.

12.3. Products and Imbeddings of Uniform Spaces.
We first mention some elementary facts.

1. If $f: X \to Y$ is a map of a set to a uniform space $(Y, \mathscr{U})$, then the set $f^{-1} \mathscr{U}$ is a basis of the smallest uniformity on X with respect to which f is uniformly continuous.

2. If $\mathscr{U}$ and $\mathscr{V}$ are uniformities on a set X, then the collection $\mathscr{U} \wedge \mathscr{V} = \{u \wedge v: u \in \mathscr{U}, v \in \mathscr{V}\}$ where $u \wedge v = \{U \cap V: U \in u, V \in v\}$ is the smallest uniformity containing $\mathscr{U}$ and $\mathscr{V}$.

3. The preceding assertion can be used to prove that there exists a fine upper bound of any collection of uniformities on a set X.

Definition. Suppose that $(X_\alpha, \mathcal{U}_\alpha)$ is a uniform space for each α in an index set A. *The product of the uniformities* is the smallest uniformity on $\prod\{X_\alpha: \alpha \in A\}$ with respect to which all projections to the coordinate spaces are uniformly continuous.

Proposition 1. *The topology of a product of uniformities coincides with the product topology.*

A map f of a uniform space to a product of uniform spaces is uniformly continuous if and only if the composition of f with each projection to a factors is uniformly continuous.

Theorem 2. *Every uniform space imbeds in a product of pseudometric spaces and every separated uniform space imbeds in a product of metric spaces.*

In the separated case, this imbedding is the diagonal product of the maps f_u in Theorem 1 with respect to all uniform covers u; in the nonseparated case, it is the diagonal product of the identity maps into the pseudometric spaces (X, d_u).

12.4. Complete Uniform Spaces. A *filter* on a set X is a collection $\mathcal{F}$ of nonempty subsets satisfying the following properties:

$F_1)$ every subset of X containing a set in $\mathcal{F}$ belongs to $\mathcal{F}$;

$F_2)$ the intersection of a finite number of sets in $\mathcal{F}$ belongs to $\mathcal{F}$.

A filter $\mathcal{F}$ on a uniform space X is called a *Cauchy filter* if $\mathcal{F} \cap u \neq \varnothing$ for any uniform cover u of X.

Definition. A uniform space X is called *complete* if every Cauchy filter $\mathcal{F}$ on X converges; that is, if there exists a point $x \in X$ each neighbourhood Ox of which belongs to $\mathcal{F}$.

Proposition 2. *Every closed subspace of a complete uniform space is complete. Conversely, every complete subspace of a separated uniform space (complete or not) is closed.*

Proposition 3. *The product of complete uniform spaces is complete. Conversely, if a product of nonempty uniform spaces is complete, then all the factor spaces are complete.*

The following theorem shows the important role of uniform spaces.

Theorem 3. *Suppose that $f: A \to Y$ is a uniformly continuous map of an everywhere dense subspace A of a uniform space X to a complete separated space Y. Then f extends to a uniformly continuous map $f: X \to Y$.*

The completion theorem for metric spaces (and a similarly-proved completion theorem for pseudometric spaces), Theorem 2, and Propositions 2 and 3 imply the following result.

Theorem 4. *Each uniform space can be imbedded in some complete uniform space as an everywhere dense subspace. Each separated uniform space imdeds as an everywhere dense subspace of a uniquely determined complete separated space, called its completion.*

Definition. A topological space $(X, \mathcal{T})$ is called *uniformly complete* if and only if there exists a uniformity $\mathcal{U}$ on X compatible with the topology $\mathcal{T}$ such that the uniform space $(X, \mathcal{U})$ is complete.

Proposition 4. *Let $(X, \mathcal{T})$ be a topological space and $\mathcal{U}$ and $\mathcal{V}$ uniformities on X compatible with the topology $\mathcal{T}$ and such that $\mathcal{U} \subset \mathcal{V}$. If $(X, \mathcal{U})$ is complete, then $(X, \mathcal{V})$ is also complete.*

Thus, it follows from Proposition 4 that a topological space X is uniformly complete if and only if it is complete in its fine uniformity. Uniform completeness of topological spaces was first investigated by Dieudonné. As a result, uniformly complete spaces are often said to be *Dieudonné complete*.

12.5. Uniformity and Compactness

Proposition 5. *The collection of all open covers of a compactum X is a basis of the unique uniformity compatible with the topology on X.*

The only difficult point here is the proof that every finite open cover $u = \{U_1, \ldots, U_n\}$ has a star refinement by a finite open cover. By the combinatorial refinement lemma (see §2, Proposition 20) a cover u can be refined to an open cover $v = \{V_1, \ldots, V_n\}$ such that $\bar{V}_i \subset U_i$. Then, upon setting $O_x = \bigcap \{V_i : x \in V_i\} \setminus \bigcup \{\bar{V}_j : x \notin V_j\}$ we obtain the desired cover $w = \{O_x : x \in X\}$.

Corollary. *Every continuous map of a compactum to a uniform space is uniformly continuous.*

Definition. A uniform space X is called *totally bounded* (or *precompact*) if every uniform cover contains a finite uniform subcover.

Proposition 6. *A uniform space is compact if and only if it is totally bounded.*

An *extension* of a topological space X is a pair (f, Y) where Y is a topological space, and $f: X \to Y$ is a homeomorphism of the space X to an everywhere dense subset of Y. On any collection of extensions of a space X one can define an order by setting $(f, Y) \geqslant (g, Z)$ if and only if there exists a continuous map $h: Y \to Z$ such that $h \circ f = g$.

Proposition 7. *The collection of all Hausdorff extensions of a space X is partially ordered by the relation $\leqslant$. If (f, y) and (g, Z) are Hausdorff extensions of X and $(f, Y) \leqslant (g, Z) \leqslant (f, Y)$, then the extensions (f, Y) and (g, Z) are topologically equivalent.*

Propositions 5 and 6 and Theorem 4 give the following.

Theorem 5. *The partially ordered set of precompact uniformities on a Tikhonov space X is isomorphic to the partially ordered set of all compact Hausdorff extensions of X. The isomorphism is realized by the operation of completion.*

12.6. Proximity Spaces and Compactifications. We consider a binary operation δ on the set $\mathscr{P}(X)$ of all subsets of a set X. If $A\delta B$, then we shall say that the sets A and B are *close*. Otherwise we say that A and B are *distant* and write $A\bar{\delta}B$. A pair (X, δ) where δ is a *proximity* on the set X will be called a *proximity space* if the relation δ satisfies the following axioms:

P_1) $A\delta B \Rightarrow B\delta A$ (the *symmetry axiom*);

P_2) $A\bar{\delta}(B_1 \cup B_2) \Leftrightarrow A\bar{\delta}B_i$, $i = 1, 2$ ((the *sum and monotonicity axiom*);

P_3) $\varnothing\bar{\delta}X$ (the *existence axiom for distant sets*);

P_4) $\{x\}\delta\{y\} \Leftrightarrow x = y$ (the *identity axiom*);

P_5) $A\bar{\delta}B \Rightarrow$ there exists C such that $A\bar{\delta}C$ and $B\bar{\delta}(X\backslash C)$ (the *normality axiom*).

Suppose that (X, δ) is a proximity space. Let $\mathscr{T}(\delta)$ be the collection of all subsets $\mathscr{T}$ of X with the property that if $x \in \mathscr{T}$, then $\{x\}\bar{\delta}(X\backslash T)$. It is easy to see that the collection $\mathscr{T}(\delta)$ is a topology on X; it is called the *topology induced by the proximity δ*.

Proposition 8. *Every topology generated by a proximity is Tikhonov.*

Indeed, the normality axiom implies that there is a dense suddivision between distant subsets of a proximity space X. In particular, there is a dense subdivision between a point and a closed set which does not contain the point. Therefore the proposition follows from Lemma 1 of § 6.

Example 1. Let (X, ρ) be a metric space. If A, B are nonempty subsets of X, set

$$A\bar{\delta}B \Leftrightarrow \rho(A, B) > 0$$

The proximity obtained this way is called a *metric proximity*. The topology it generates on X is the same as that generated by the metric ρ. Different metrics can generate the same proximity.

Example 2. Let $(X, \mathscr{U})$ be a separated uniform space. Set $A\bar{\delta}_{\mathscr{U}}B \Leftrightarrow St_u A \cap St_u B = \varnothing$ for some $u \in \mathscr{U}$. We obtain a proximity called the *uniform proximity* or *proximity generated by the uniformity*. It can be shown that any proximity is generated by some uniformity.

The collection of all proximities on X, being a collection of subsets of the product $\mathscr{P}(X) \times \mathscr{P}(X)$, is ordered by inclusion. We shall say that a proximity δ_1 is *larger* than a proximity δ_2 if $A\delta_1 B$ implies $A\delta_2 B$.

Example 3. Let $(X, \mathscr{T})$ be a Tikhonov space. Declare any nonempty sets A and B to be distant if they are functionally separated; that is, if there exists a continuous function $\varphi: X \to [0, 1]$ for which $\varphi(A) = 0$ and $\varphi(B) = 1$. The resulting proximity is the largest proximity generating the given topology. This follows because any two nonempty distant sets are functionally separated (as shown in the proof of Proposition 8).

Example 4. On an arbitrary set X we can put

$$A \bar{\delta} B \Leftrightarrow A \cap B = \varnothing.$$

The proximity relation obtained thereby is the largest proximity on the set X. It generates the discrete topology on X.

Proposition 9. *Let (X, δ) be a proximity space. Among the uniformities generating δ (see Example 2), there is a smallest uniformity $\mathcal{U}_\delta$. This uniformity is precompact.*

Theorem 6. Let $(X, \mathcal{T})$ be a Tikhonov space. Then the mapping $\mathcal{U} \to \delta_{\mathcal{U}}$ is an isomorphism between the partially ordered set $U_{pc}(\mathcal{T})$ of precompact uniformities generating the topology $\mathcal{T}$ and the partially ordered set $P(\mathcal{T})$ of all proximities generating $\mathcal{T}$.

From Theorems 5 and 6, we obtain the following.

Theorem 7. *The partially ordered set of all Hausdorff compactifications of a Tikhonov space X is isomorphic to the partially ordered set of all proximities compatible with the space X. This isomorphism is realized by the operation which associates to a compactification cX the proximity δ_c on X defined by:*

$$A \delta B \Leftrightarrow A^{cX} \cap B^{cX} = \varnothing$$

Comments on the Bibliography

The first steps towards the founding of general topology appear in Frechet's memoir [Fre] and Hausdorff's book [H1]. The fundamental idea of compactness appears in explicit form in the memoir of Aleksandrov and Uryson [AU].

Among the textbooks, the most complete accounts of general topology are to be found in Engelking's book *General Topology* [E] and Arkhangel'skiĭ and Ponomarev's book [AP]. Other books which would serve as academic textbooks are those of Aleksandrov and Uryson [AU], Kelley [Ke], Kuratowski ([Ku1] and [Ku2]), Nagata [N], Bourbaki ([B3], [B2] and [B1]), and Franklin ([F1] and [F2]). A popular exposition of the ideas of general topology is set forth in the books by Gähler [G] and Čech [C2]. The survey articles of Aleksandrov ([A4] and [A6]) and Arkhangel'skiĭ [Ar3] sketch the state of different branches of general topology at a particular stage in their development. The articles by Aleksandrov ([A1], [A2] and [A3]), Arkhangel'skiĭ [Ar3], Shchepin [Sh], Čech [C1], Michael [M2], Tikhonov ([T2] and [T3]), A.N. Stone ([St1], [St2]), and M. H. Stone [S2] are among those that played an important role in the development of general topology.

More detailed accounts of the different areas of general topology, the theory of cardinal invariants, dimension theory, uniform topology, and so on, as well as other areas closely connected with general topology can be found in the monographs by Aleksandrov and Pasynkov [AP], Naimark [Na], Pontryagin ([P1] and [P2]), Isbell [I], Rudin [R], Walker [Wa], Weil [W], and Whyburn [Wh].

Bibliography*

[A1]   Alexandroff, P.S. (= Alexandrov, P.S.): Sur les ensembles de la première classe et les espaces abstraits. C.R. Acad. Sci., Pairs *178*, No. 3 (1924) 185–187. Jrb.50,134

[A2]   Alexandroff, P.S.: Uber stetige Abbildungen kompakter Räume. Math. Ann. *96*, No. 3 (1926) 555–571. Jrb.52,584

[A3]   Aleksandrov, P.S.: On bicompact extensions of topological spaces, Mat. Sb., Nov. Ser. *5*, No. 4 (1939) 403–423 (Russian). Zbl.22,412

[A4]   Aleksandrov, P.S.: On the concept of space in topology, Usp. Mat. Nauk *2*, No. 1 (1947) 5–57 (Russian)

[A5]   Aleksandrov, P.S.: Combinatorial Topology, Gostekhizdat, Moscow, 1947. 660pp. (Russian). Zbl.37,97

[A6]   Aleksandrov, P.S.: Introduction to Set Theory and and General Topology, Nauka, Moscow, 1977. 367 pp. (Russian)

[AF]   Aleksandrov, P.S., Fedorchuk, V.V.: Fundamental aspects in the development of set-theoretical topology, Usp. Mat. Nauk *33*, No. 3 (1978) 3–48. Zbl.389.54001; English translation: Russ. Math. Surv. 33, No. 3 (1978) 1–53

[AH]   Alexandroff, P.S., Hopf, H.: Topologie 1, Berlin, 1935. 636 pp. Zbl.13,79

[AP]   Aleksandrov, P.S., Pasynkov, B.A.: Introduction to Dimension Theory, Nauka, Moscow, 1973. 575 pp. (Russian). Zbl.272.54028

[AU]   Alexandroff, P.S., Urysohn, P.S. (= Uryson, P.S.), Mémoire sur les espaces topologiques compacts. Verh. Acad. Wet. Amsterdam *14* (1929) 93 pp. Jrb.55,960

[ArF]   Arhangel'ski, A.V. (= Arkhangel'skiĭ, A.V.), Franklin, S.P.: Ordinal invariants for topological spaces, Mich. Math. J. *5* (1968) 313–320. Zbl.167,511

[Ar1]   Arkhangel'skiĭ, A.V.: On maps of metric spaces, Dokl. Akad. Nauk SSSR *145*, No 2 (1962) 245–247. Zbl.124,158; English translation: Sov. Math., Dokl. 3 (1962) 953–956

[Ar2]   Arkhangel'skiĭ, A.V.: Some types of quotient mappings and connections between classes of topological spaces, Dokl. Akad. Nauk SSSR *153*, No. 4 (1963) 743–746. Zbl.129,381; English translation: Sov. Math., Dokl. 4 (1963) 1726–1729

[Ar3]   Arkhangel'skiĭ, A.V.: Mappings and spaces, Usp. Mat. Nauk *21*, No. 4 (1966) 133–184. Zbl.171,436; English translation: Russ. Math. Surv. 21, No. 4 (1966) 115–162

[Ar4]   Arkhangel'skiĭ, A.V.: Intersections of topologies and pseudo-open bicompact maps, Dokl. Akad. Nauk SSSR *226*. No. 4 (1976) 745–748. Zbl.344.54010; English translation: Sov. Math., Dokl. 17, (1976) 160–163

[Ar5]   Arkhangel'skiĭ A.V.: Construction and classification of topological spaces and cardinal invariants, Usp. Mat. Nauk *33*, No. 6 (1978) 29–84. Zbl.414.54002; English translation: Russ. Math. Surv. 33, No. 6 (1978) 33–96

[ArP]   Arkhangel'skiĭ, A.V., Ponomarev V.I.: The Principles of General Topology through Problems and Excerises, Nauka, Moscow, 1974. 424 pp. (Russian) Zbl.287.54001

[Ba1]   Baire, R.: Sur la représentation des fonctions discontinues. Acta. Math. *30* (1905). 1–48 Jrb.36,453

[Ba2]   Baire, R.: Leçons sur les fonctions discontinues, Gauthier – Villars, Pairs, 1905. 127 pp. Jrb.36,438

[Ba3]   Baire, R.: Theory of Discontinuous Functions, GTTI, Moscow-Leningrad, 1932. 136 pp. (Russian)

[B1]   Bourbaki, N.: Topologie générale, Ch. X, Pairs, 1949. Zbl.36,386

[B2]   Bourbaki, N.: Topologie générale, Ch. IX (2ème ed.), Pairs, 1958, 169 pp. Zbl.85,371

[B3]   Bourbaki, N.: Topologie générale, Ch. 1 et 2 (3éme ed.), Paris, 1961, 263 pp. Zbl.102,376

---

*For the convenience of the reader, references to reviews in Zentralblatt für Mathematik (Zbl.), compiled using the MATH database, and Jahrbuch über die Fortschritte der Mathematik (Jrb.) have, as far as possible, been included in this bibliography.

[B4]   Bourbaki, N.: Algèbre commutative, Ch 1 et 2., Hermann, Paris, 1969, 578 pp. Zbl.108,40
[C1]   Čech, E.: On bicompact spaces, Ann. Math., II. Ser. *38* (1937) 823–844. Zbl.17,428
[C2]   Čech, E.: Topological spaces, Acad., Prague, 1966, 479 pp. Zbl.141,394
[Co]   Cohen, D. E.: Spaces with weak topology, Q. J. Math., Oxf. II. Ser. *5*, No. 17 (1954), 77–80.
          Zbl.55,161
[E]   Engelking, R.: General topology, PWN, Warszawa, 1977, 626 pp. Zbl.373.54002
[F1]   Fedorchuk, V.V.: Products and Spectra of Topological Spaces, Part I, Izd-vo MGU,
          Moscow, 1979, 87 pp. (Russian)
[F2]   Fedorchuk, V.V.: Products and Spectra of Topological Spaces, Part II, Izd-vo MGU,
          Moscow, 1980, 94 pp. (Russian). Zbl.536.54001
[Fi]   Filippov, V.V.: Quotient spaces and the multiplicity of a basis, Mat. Sb., Nov. Ser. *80*, No.
          4 (1969) 521–532. Zbl.199,573; English translation; Math. USSR, Sb. 9 (1969) 487–496
[Fr1]   Franklin, S.P.: Spaces in which sequences suffice, Fundam. Math. *57* (1965) 107–115.
          Zbl.132,178
[Fr2]   Franklin, S.P.: Spaces in which sequences suffice 2. Fundam. Math. *61* (1967) 51–56.
          Zbl.168,435
[Fre]   Fréchet, M.: Sur quelques points du calcul fonctionnel, Rend. Circ. Mat. Palermo *22* (1906)
          1–74. Jrb.37,348
[G]   Gähler, W.: Grundstrukturen der Analysis 1, 2, Berlin/Basel-Stuttgart, 1977 (Part 1, 12 pp.)
          Zbl.346.54001, 1978 (Part 2, 623 pp.)Zbl.393.46001
[H1]   Hausdorff, F.: Grundzüge der Mengenlehre, Leipzig, 1914, 476 pp.
[H2]   Hausdorff, F.: Mengenlehre, Berlin, 1927, 285 pp.
[HW]   Hurewicz, W., Wallman, H.: Dimension theory, Princeton, 1948, 165 pp. Zbl.36,125
[I]   Isbell, J.R.: Uniform spaces. Providence: Rhode Island, 1964, 175 pp. Zbl.124,156
[K]   Keldysh, L.V.: Zero-dimensional open maps, Izv. Akad. Nauk SSSR, Ser. Mat. *23*, No. 2
          (1959) 165–184. (Russian). Zbl.86,158
[Ke]   Kelley, J.L.: General topology, Van Nostrand, Princeton & New York, 1955, 298 pp.
          Zbl.66,166
[Ko]   Kowalsky, H.J.: Topologische Räume, Birkhäuser, Basel-Stuttgart, 1961, 271 pp. Zbl.93,361
[Ku1]   Kuratowski, K.: Topology, Vol. 1, New York, 1966, 560 pp. Zbl.158,408
[Ku2]   Kuratowski, K.: Topology, Vol. 2, New York, 1968, 511 pp. Zbl.102,376
[L]   Lashnev, N.S.: On continuous partitions and closed maps of metric spaces, Dokl. Akad.
          Nauk SSSR *165*, No. 4 (1965) 756–758. Zbl.145,196; English translation: Sov. Math., Dokl.
          6 (1965) 1504–1506
[La]   Lavrentjeff, M.A. (= Lavrent'ev M.A.): Contribution à la théorie des ensembles homéo-
          morphes, Fundam. Math. *6* (1924) 149–160. Jrb.50,143
[M1]   Michael, E.: On representing spaces as images of metrizable and related spaces, Gen.
          Topology Appl. *1*, No. 4 (1971) 329–343. Zbl.272.54009
[M2]   Michael, E.: A quintuple quotient quest, Gen. Topology Appl. 2, No. 2 (1972) 91–138.
          Zbl.238.54009
[M3]   Michael, E.: On k-spaces, k_R-spaces and $k(X)$, Pac. J. Math. *47*, No. 2 (1973) 487–498.
          Zbl.225.54022
[My]   Mysior, A.: A regular space which is not completely regular, Proc. Am. Math. Soc. *81*, No.
          4 (1981) 52–53. Zbl.451.54019
[N]   Nagata, J.: Modern General Topology, North-Holland, Amsterdam, 1968, 353 pp. Zbl.181,
          254
[Na]   Naĭmark, M.A. (= Najmark, M.A.): Normed Rings, Nauka, Moscow, 1968 (Second edition),
          664 pp. (Russian). Zbl.175,437
[Nel]   Nedev, S.: Symmetrizable spaces and final compactness, Dokl. Akad. Nauk SSSR *175*, No.
          3 (1967) 532–534. Zbl.153,527; English translation; Sov. Math., Dokl. 8 (1967) 890–892
[Ne2]   Nedev, S.: o-metrizable spaces, *Tr. Mosk. Mat. o.-va. 24* (1971) 201–236. Zbl.244.54016;
          English translation: Trans. Mosc. Math. Soc. 24 (1974) 213–247
[No]   Noble, N.: Countably compact and pseudocompact products, Czech. Math. J. *19*, No. 3
          (1969) 390–397. Zbl.184,477

[Nv]   Novák, J.: A regular space on which every continuous function is constant, Čas. Mat. Fys. *73* (1948) 58–68. Zbl. 32,431

[Ox]   Oxtoby, J.C.: Measure and Category. A Survey of the Analogies between Topological and Measure Spaces, Springer (Grad. Text. Math. 2), Berlin & New York. (1971) 95 pp. Zbl.217,92

[PF]   Pasynkov, B.A., Fedorchuk, V.V.: Topology and Dimension Theory, Znaniye (Ser. Mat. Kibern. No. 9), Moscow, 1984, 64 pp. (Russian)

[Po]   Ponomarev, V.I.: Axioms of countability and continuous maps, Bull. Acad. Pol. Sci., Sér. Sci. Math. Astron. Phys. *8*, No. 3 (1960) 127–134 (Russian). Zbl.95,163

[P1]   Pontryagin, L.S.: Fundamentals of Combinatorial Topology, Nauka, Moscow, 1976 (second edition), 136 pp. (Russian). Zbl.463.55001

[P2]   Pontryagin, L.S.: Continuous Groups, Fizmatgiz, Moscow, 1954, 515 pp (Russian). Zbl.58, 260

[R]   Rudin, M.E.: Lectures on Set Theoretic Topology, Providence, 1975, 76 pp. Zbl.318.54001

[Ru]   Rudin, W.: Functional analysis, McGraw-Hill Book Company, New York & Toronto, 1973, 397 pp. Zbl.253,46001

[Sto]   Stoilow, S.: Sur les transformation continues et la topologie des fonctions analytiques, Ann. Sci. Ecole Norm. Supér., III. Sér. *45* (1928) 347–382. Jrb.54,607

[Sh]   Shchepin, E.V.: The topology of limit spaces of uncountable inverse spectra, Usp. Mat. Nauk *31*, No. 5 (1976) 191–226. Zbl.345.54022; English translation; Russ. Math. Surv. 31, No. 5 (1876) 155–191

[St1]   Stone, A.H.: Paracompactness and product spaces, Bull. Am. Math. Soc. *54*, No. 4 (1948) 977–982. Zbl.32,314

[St2]   Stone, A.H.: Metrizability of decomposition spaces, Proc. Am. Math. Soc. 7, No. 3 (1956) 690–700. Zbl.71,160

[S1]   Stone, M.H.: Applications of Boolean algebras to topology. Mat. Sb., Nov. Ser. *1*, No. 6 (1936) 765–771. Zbl.16,182

[S2]   Stone, M.H.: Applications of the theory of Boolean rings to general topology, Trans. Am. Math. Soc. *41*, No. 2 (1937) 375–481. Zbl.17,135

[S3]   Stone, M.H.: The generalized Weierstrass approximation theorem, Math. Mag. *21* (1948) 167–184, 237–254

[TF]   Terpe, F., Flachsmeyer, J.: On some applications of the theory of extensions of topological spaces and measure theory, Usp. Mat. Nauk *32*, No. 5 (1977) 125–162. Zbl.374.54017; English translation: Russ. Math. Surv. 32, No. 5 (1977) 133–171

[Tu]   Tukey, J.W.: Convergence and Uniformity in Topology, Ann. Math. Stud. 2, Princeton, 1940, 90 pp. Zbl.25,91

[T1]   Tychonoff, A.N. (= Tikhonov A.N.): Über einen Metrisationssatz von P. Urysohn, Math. Ann. *95* (1925) 139–142. Jrb.51,453

[T2]   Tychonoff, A.N.: Über die topologische Erweiterung von Räumen, Math. Ann. *102* (1929) 544–561. Jrb.55,963

[T3]   Tychonoff, A.N.: Über einen Funktionenraum, Math. Ann. *111* (1935) 762–766. Zbl.12,308

[U]   Urysohn, P.S. (= Uryson, P.S.): Sur la métrisation des espaces topologiques. Bull. Int. Acad. Pol. Sci., Sér. A, (1923) 13–16. Jrb.49,405

[Us]   Uspenskiĭ, V.V.: Pseudocompact spaces with σ-point-finite base are metrizable, Comment. Math. Univ. Carol. *25*, No. 2 (1984) 261–264. Zbl.574.54021

[Wa]   Walker, R.C.: The Stone-Čech Compactification, Springer-Verlag, 1974, 332 pp. Zbl.292. 54001

[Wat]   Watson W.S.: Pseudocompact metacompact spaces are compact, Proc. Am. Math. Soc. *81*, No. 1 (1981) 151–152. Zbl.468.54014

[W]   Weil, A.: Sur les espaces à structure uniforme et la topologie generale, Paris, 1938, 40 pp. Zbl.19,186

[Wh]   Whyburn, G.T.: Analytic Topology, New York, 1942, 278 pp. Zbl.36,124

II. The Fundamentals of Dimension Theory

V.V. Fedorchuk

Translated from the Russian
by D. O'Shea

Contents

Introduction

1. Brief Historical Sketch. Together with the theory of continua, dimension theory is the oldest branch of general topology. The first concepts and facts predate Hausdorff's definition in 1914 of general Hausdorff topological spaces and, so, involved only subsets of Euclidean spaces. In its infancy, dimension theory was nurtured by the work of three outstanding mathematicians: Poincaré, Brouwer, and Lebesgue. Peano's construction in 1890 of a continuous map of a segment onto a square gave rise to the problem of whether the dimension of Euclidean space was a topological invariant. This problem was solved by Brouwer in 1911 (see the article [B1]) using the concept he introduced of the degree of a map. In the same paper Brouwer proved that, for $\varepsilon < \frac{1}{2}$, the Euclidean cube I^n could not be mapped by an ε-shift to a nowhere dense subset A of I^n (we now know that such a set A has dimension less than n).

Another important step towards the definition of dimension was also taken in 1911. Lebesgue formulated the covering theorem for the n-dimensional cube I^n and attempted to prove it in [L1]. Thus, Lebesgue was the first mathematician to understand the connection between dimension and multiplicity of covers. The first irreproachable proof of Lebesgue's theorem was given by Brouwer in 1913 in the paper [B2]; Lebesgue, himself, proved his theorem only in 1921 in [L2].

In 1912 an expository paper [Po] by Poincaré appeared in which he raised the deep possibility of inductively defining the number of dimensions of a space by considering the sets of lower dimension which separate it. In 1913, Poincaré's idea received a precise mathematical treatment in the aforementioned paper [B2]. In this paper, Brouwer introduced a dimensional invariant which, for complete metric spaces, coincides with the large inductive dimension Ind and proved the equality Ind $\mathbb{R}^n = n$.

However, it only became possible to speak of dimension theory as an independent domain of general topology after the work of Menger and Uryson. In 1921 they arrived, independently of Brouwer and one another, at the concept of the small inductive dimension ind and used it as a basis for the systematic construction of dimension theory. Their first results were the decomposition theorem for separable metric spaces into zero-dimensional sets and the countable sum theorem for metric compacta. In addition, Uryson introduced the dimension dim as an independent topological invariant and proved that if X was a metric compactum, then

$$\dim X = \operatorname{ind} X = \operatorname{Ind} X.$$

In his principal work on dimension theory, *Memoir on Cantor Manifolds*, Uryson introduced and investigated the concept of an n-dimensional Cantor manifold. He considered this concept to be one of the central concepts in dimension theory and stated the two following conjectures:

1) Every n-dimensional metric compactum contains an n-dimensional Cantor manifold.

2) The common boundary of a bounded domain and its complement in n-dimensional Euclidean space is an $(n - 1)$-dimensional Cantor manifold.

Hurewicz and Tumarkin proved the first conjecture in 1926. Uryson, himself, proved the second conjecture in the case $n = 3$. In 1927 Aleksandrov established the general case as a precursor of the general homological theory of dimension.

An important step in geometrizing the theory of dimension was the theorem Aleksandrov proved in 1926 that, for any $\varepsilon > 0$, an n-dimensional compactum X lying in a Hilbert or Euclidean space R can be transformed into a polyhedron of dimension n (and no smaller dimension) by means of an ε-shift $f: X \to R$.

After the foundations of the general theory of dimension had been laid in the works of Uryson and Menger, subsequent development (apart from the aforementioned results of Aleksandrov) was connected with the extension by Hurewicz and Tumarkin in 1925–1926 of the dimension theory of metric compacta to metric spaces with a countable base and with the theorems of Hurewicz about the dimension of mappings (see Chap. 6).

One of the most significant results in the theory of dimension was obtained in 1930–1931: this is the Nöbeling-Pontryagin theorem which states that any n-dimensional metric compactum is homeomorphic to a subspace of $(2n + 1)$-dimensional Euclidean space. Taken together with Hurewicz's theorem that every n-dimensional, separable metric space lies in some n-dimensional metric compactum and Uryson's metrization theorem, the Nöbeling-Pontryagin theorem gives the following fundamental fact: any finite-dimensional normal space with a countable base is (and only these are) homeomorphic to a subspace of Euclidean space.

In 1932, Aleksandrov set forth the basis of the homological theory of dimension [A1]. A precursor of this theory was the theorem, cited above, about the common boundary of two domains. Another was the theorem about the existence of essential maps of n-dimensional spaces to the n-dimensional cube which Aleksandrov proved in 1930 using the theorem about approximation of n-dimensional spaces by n-dimensional polyhedra. Many subsequent developments of the homological theory of dimension were connected with Pontryagin's work. Bokshtein, Boltyanskiĭ, Borsuk, Sitnikov, Dyer, Kodama, Kuz'minov, and others also contributed.

In the 1930's the dimension theory of general spaces was developed. The first steps were made by Čech, who proved the monotonicity theorem for the dimensions Ind and dim, the finite sum theorem for the dimension Ind of perfectly normal spaces, and the countable sum theorem for the dimension dim. Another important step was taken with the introduction by Kuratowski of barycentric mappings and his proof of the existence of ω-maps of any normal space to the body of the nerve of an arbitrary finite open cover ω. Aleksandrov subsequently characterized the dimension of normal spaces by means of ω-maps to polyhedra. It turned out to be possible to reduce the study of certain dimensional properties of general spaces to the study of dimensional properties of spaces with a countable base and even polyhedra.

The 1940's and 1950's saw an intensive study of the dimensional properties of paracompact and metric spaces. Especially noteworthy is the work of Dowker,

Katětov, Morita, Nagata, and Smirnov. By the mid 60's, the dimension theory of metric spaces was complete.

Concerning the dimensions of general spaces, one has the results of Dowker about the dimension Ind of hereditarily normal spaces. Sklyarenko, Pasynkov, Zarelya proved theorems about compactifications, universal spaces, and dimensions of maps. Examples due to Lunts, Lokutsievskiĭ, Fedorchuk, and Filippov showed that the dimensional properties of nonmetrizable compact spaces are not much better than those of more general spaces.

Particularly interesting from the point of view of dimension theory are open maps. Whereas zero-dimensional spaces can be mapped onto spaces of arbitrarily large dimension by closed finite-to-one maps, a countable-to-one open mapping of a compactum cannot raise its dimension. This was established by Aleksandrov (1936) for metric compacta and Pasynkov (1967) in the general case.

In contrast, arbitrary (even zero-dimensional) open mappings can raise dimension. The first example of an open and zero-dimensional map from a metrizable compactum to a compactum of greater dimension was constructed in 1936 by Kolmogorov. Many other examples of open mappings which raise dimension were subsequently found. Relying on Keldysh's example of an open zero-dimensional map of a one-dimensional compactum to a square, Pasynkov was able to prove that any compactum of positive dimension is the image of a one-dimensional map. Very recently (1983), within the framework of the theory of extensors, A.N. Dranishnikov obtained very precise results about open mappings which raise dimension.

One of the more significant events in general topology has been the birth and rapid development of the theory of infinite dimensional spaces. The basic concepts of this fascinating chapter of dimension theory and some of the principal problems associated with it were already known to Uryson. But he did not succeed in proving a single one of his conjectures connected with infinite dimensional spaces and never published anything about them. The basis of the theory of infinite-dimensional spaces was set out significantly later in the book of Hurewicz and Wallman [HW] which was published in 1941. Large contributions to the theory were subsequently made by Smirnov, Nagata, Sklyarenko, Levshenko, Pasynkov, Fedorchuk, Luxemburg, R. Pol, and others.

2. General Remarks. This article is neither a handbook nor a textbook on dimension theory. The restricted length does not allow us to even survey the main results of so comprehensively developed an area of topology. In fact, this article omits almost all mention of the following topics: dimension and the problem of extension of maps, dimension of spaces in the presence of algebraic structures, dimension and covariant functors. The same remarks also apply to the short historical sketch, which could not include all the names and results which merit mention. Many additional historical facts are scattered throughout the article.

Although the choice of material cannot help but reflect the mathematical tastes and interests of the author, I hope that those concepts and facts of dimension theory of greatest common interest are treated in adequate detail. Many assertions are provided with sketches of proofs. This allows us, on the one hand, to cover the main ideas and methods of dimension theory and, on the other, to introduce a pedagogical element which will allow the diligent reader to fill in the gaps.

All spaces are supposed Hausdorff and, as a rule, normal; all maps, unless the contrary is explicity stated, are continuous.

Chapter 1
The Principal Definitions and Simplest Facts of Dimension Theory

§ 1. What is a Curve?

1.1. The Earliest Definitions of a Curve. Jordan Curves. The Peano Curve. The concept of a curve is at once one of the principal objects and one of the principal tools of geometrical investigation. It arose in practical work and finds wide application to the mathematical description of natural and man-made processes.

Attempts to precisely define curves stem from the very beginnings of mathematics. In the "Elements", Euclid defines a line to be that which has "length but no breadth" or that which "bounds a surface". Within the framework of elementary geometry it is difficult to give a more precise definition and, hence, at that level the study of curves reduces to considering particular examples: lines, segments, broken lines, circles, and so forth. The systematic application of coordinate methods, begun by Fermat and Descartes, significantly enriched geometry by allowing the use of functional methods in geometric investigations. Curves in the plane were defined as sets of points whose coordinates satisfy an equation

$$F(x, y) = 0 \tag{1}$$

But even if F is required to be differentiable, every closed subset of the plane satisfies this definition. Consideration of algebraic curves (that is, subsets of the plane satisfying (1) with $F(x, y)$ a polynomial in two variables) leads to an important field of mathematics, namely algebraic geometry, but is much too restrictive for geometry, not to mention other areas of mathematics.

The specification of curves by parametric equations occurs in many branches of mathematics. Here, for example, a curve in the plane given by parametric equations

$$x = \varphi(t), y = \psi(t), \tag{2}$$

where φ and ψ are continuous (and, as a rule, differentiable) functions defined on an interval $a \leqslant t \leqslant b$, is the locus of points (x, y) corresponding to all values of the parameter t. One frequently assumes, for convenience, that points corresponding to different values of the parameter are different.

However, this definition of a curve is not explicitly topological, and the rise of topology in the second half of the nineteenth century required a topological definition. One attempt at such a definition was made by Jordan in 1882. He proposed that a curve be defined as a continuous image of an interval. In the case of the plane, this is just the set of points given by the parametric equations (2); but, φ and ψ are now arbitrary continuous functions and points corresponding to different parameter values which have the same coordinates are no longer considered distinct. This definition carries over to any topological space: a set of points of a topological space which is the continuous image of an interval is said to be a *Jordan curve*.

In 1890 the Italian mathematician, Peano, dealt a blow to this definition of a curve by constructing a continuous mapping of an interval onto a square, a so-called *Peano curve*. We cite a construction of a Peano curve due to Hilbert. The idea of the construction is as follows. At the nth step, the interval I and the square I^2 are divided into 4^n identical intervals ^nI_i and squares $^nI_i^2$, respectively. The intervals are enumerated in the natural order from left to right. The squares are enumerated so that

 a) $^nI_i^2$ and $^nI_{i+1}^2$ have a common side;

 b) if $^{n+1}I_{i_k}^2 \subset {}^nI_{j_k}^2$, $k = 1, 2$ and $j_1 < j_2$ then $i_1 < i_2$.

Define a map $f\colon I \to I^2$ as follows. If $x = \bigcap \{^nI_{i_n}\colon i = 1, 2, \ldots\}$, then $f(x) = \bigcap \{^nI_{i_n}^2\colon n = 1, 2, \ldots\}$. Conditions a) and b) guarantee that f is well-defined and continuous. The first three steps of the construction are depicted in Fig. 1. The direction of the circuit about the square corresponds to the order of enumeration.

This construction possesses the following property. For each point $y \in I^2$ there exists a number $k(y)$ such that one of the following conditions holds for all $n \geqslant k(y)$:

 O_1) y belongs to exactly one square $^nI_i^2$;

 O_2) y belongs to exactly two squares $^nI_i^2$;

 O_4) y belongs to exactly four squares $^nI_i^2$.

It is easy to see that if condition O_m) holds, the inverse image $f^{-1}y$ of y consists of m points. Thus, the map $f\colon I \to I^2$ has quadruple points. The curve which Peano actually constructed has at worst triple points. One can show that there

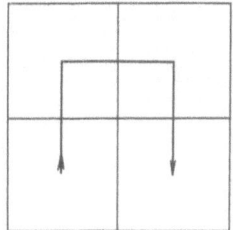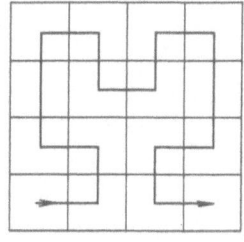

Fig. 1

is no Peano curve with at worst double points. In fact, it is impossible to map a segment countinuously onto a set of dimension greater than or equal to two in such a way that there are at worst double points. Closely connected with Peano curves is the curious fact that there exist simple arcs (that is, homeomorphic images of the interval) in three dimensional Euclidean space which project onto the square – an example is the curve $x = \varphi(t)$, $y = \psi(t)$, $z = t$, where the first two functions define a Peano curve. This curve is called a *Peano umbrella* and is "impermeable to vertical rain". Even more striking is the fact that there also exist zero-dimensional umbrellas over the square (this follows from Theorem 5 of Chapter 6).

Jordan curves (that is, continuous images of a segment) are nowadays called Peano continua. The existence of a Peano curve immediately implies the existence of continuous maps of an interval onto a cube of any finite number of dimensions. In fact, the following holds.

The Hahn-Mazurkiewicz Theorem. *The Peano continua are precisely the locally connected continua.*

1.2. Cantor Curves. The Sierpinski Carpet. Jordan's definition of a curve was unsatisfactory not only because it admitted objects which were quite unlike curves; but because there exist continua which can naturally be considered as curves, but which are not locally connected and, hence, not continuous images of an interval. An example is the continuum consisting of the graph of the function

$$y = \sin(1/x), \qquad 0 < x \leqslant 1,$$

together with the limiting segment

$$x = 0, \qquad -1 \leqslant y \leqslant 1 \qquad \text{(Fig. 2)}.$$

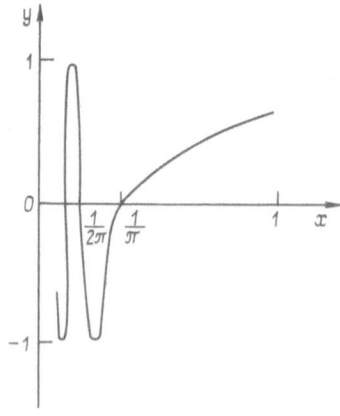

Fig. 2

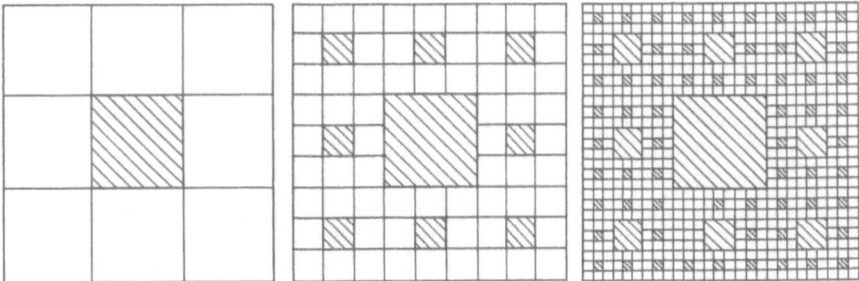

Fig. 3

In 1870, Cantor gave a general definition of a curve on the plane. A *Cantor curve* is a plane continuum which does not contain interior points with respect to the plane; that is, it is a continuum each point of which has the property that any neighbourhood contains points of the plane which do not belong to the continuum. The only drawback was that this definition is formulated in terms of the disposition of the continuum on the plane and it is not clear whether it is topologically invariant. There was no guarantee that a continuum might have interior points for one embedding into the plane, but none for another. This drawback disappeared in 1912 after Brouwer proved the invariance theorem for interior points of subsets of Euclidean space (see Chap. 2, § 2).

The most important example of a Cantor curve is the *Sierpiński carpet* which is constructed as follows. Divide the unit square I^2 into nine congruent squares by lines parallel to the sides and remove the interior of the center square. Perform the same procedure on each of the remaining eight squares, called squares of the first rank, to obtain 64 squares of the second rank. Iterate this procedure. At the nth step, we obtain 8^n squares of the nth rank with sides of length $(1/3)^n$. The intersection of the sets obtained in this manner is called the Sierpiński carpet S. The first three steps of the process are depicted in Fig. 3 (the shaded squares are the ones removed).

The Sierpiński carpet is a locally connected continuum and, consequently, must be the continuous image of an interval by the Hahn-Mazurkiewicz theorem. It is *universal* for Cantor curves. This means that any Cantor curve L can be topologically embedded in the Sierpinski carpet S; that is, S contains a continuum L' homeomorphic to L. Moreover, the homeomorphism $h: L \to L'$ can be extended to a homeomorphism $\bar{h}: \mathbb{R}^2 \to \mathbb{R}^{2'}$ of the plane to itself. Figure 4 indicates the idea for constructing the homeomorphism $\bar{h}$.

The standard Sierpiński carpet S is obtained from the Euclidean square on the plane by throwing out a set of full measure; that is, the measure of the Sierpinski carpet S is equal to zero (the measure of the set remaining after the nth step is $(8/9)^n$). It is possible to construct Sierpinski carpets with measure arbitrarily close to 1. Moreover, the following holds.

Theorem 1. *Let B_n be closed sets in the interior of the square I^2 which are homeomorphic to closed disks and which satisfy the conditions:*

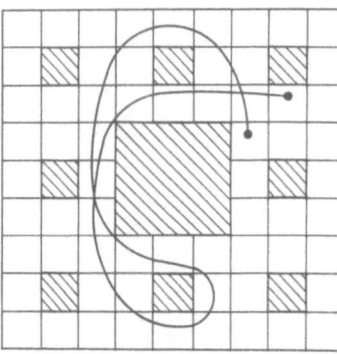

Fig. 4

1) $B_{n_1} \cap B_{n_2} = \varnothing$,

2) $\bigcup \{B_n : n = 1, 2, \ldots\}$ *is everywhere dense in* I^2,

3) diam $B_n \to 0$ *as* $n \to \infty$.

Then the set $I^2 \setminus \bigcup \{\text{Int } B_n : n = 1, 2, \ldots\}$ *is homeomorphic to a Sierpiński carpet.*

1.3. Uryson's Definition of a Line. Menger's Curve. In 1921, Uryson gave a general, topologically invariant definition of a curve. He defined a *curve* to be a *one-dimensional continuum*; that is, a connected compact metrizable space, each point of which possesses an arbitrarily small neighbourhood with zero-dimensional boundary.

The Sierpinski carpet satisfies Uryson's definition of a curve. Indeed, any vertical or horizontal straight line passing through the center intersects the Sierpinski carpet in a set homeomorphic to a Cantor perfect set (see § 3). Thus each point of the Sierpinski carpet possesses an arbitrarily small "rectangular", and even "square", neighbourhood whose boundary is homeomorphic to a Cantor perfect set and, hence, zero-dimensional.

It follows that each Cantor curve, being homeomorphic to a subset of the Sierpinski carpet, is also one-dimensional and, consequently, a curve in the sense of Uryson. Conversely, a one-dimensional plane continuum is a Cantor curve. Indeed, a one-dimensional dense set cannot have interior points with respect to the plane because each nonempty open subset of the plane is two-dimensional (see Chap. 2, § 7).

There exist curves which are not homeomorphic to any subset of the plane. An example is the curve in three-dimensional space consisting of the six edges of a tetrahedron and the segment joining the midpoints of one pair of non-intersecting edges.

Menger's Theorem. *Every curve is homeomorphic to a subset of three-dimensional Euclidean space.*

This theorem and the more general Nöbeling-Pontryagin theorem will be dealt with in more detail in Chapter 2, § 4. Menger also obtained a stronger result

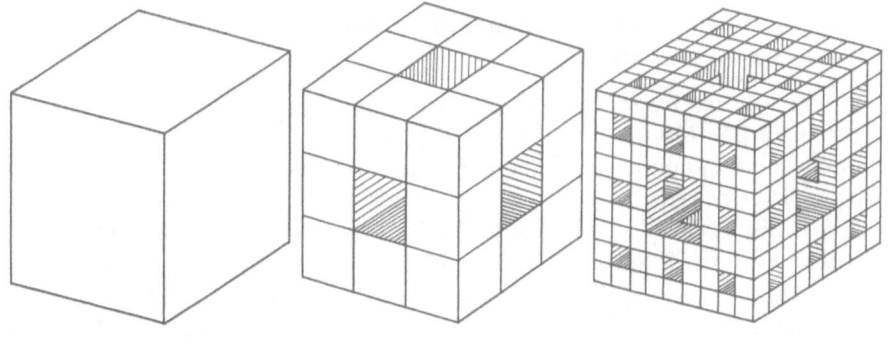

Fig. 5

(1926). He constructed a one-dimensional continnum M_1^3 in $\mathbb{R}^3$ which contains a homeomorphic image of every curve. This continuum is called the *universal Menger curve*.

The universal Menger curve M_1^3 is constructed as follows. Divide the unit cube I^3 by planes parallel to the faces into 27 congruent cubes with sides of length 1/3, called cubes of the first rank. Delete the center cube and the six cubes adjacent to it. Let K_1 denote the closure of the set that remains. The continuum K_1 can also be described as follows. It is the union of the twenty cubes of the first rank which have points in common with the one-skeleton (that is, the union of the edges) of the original cube. We carry out the same procedure on each cube of the first rank to obtain a continuum K_2 consisting of 400 cubes of the second rank. We continue this process to obtain a decreasing sequence of continua $I^3 = K_0 \supset K_1 \supset K_2 \supset \cdots$ whose intersection is the continuum M_1^3. The first two steps of this construction are depicted in Fig. 5.

Each plane parallel to a coordinate plane and passing through the center of an inside cube of first rank intersects the continuum M_1^3 in a set which is the cartesian product of a Cantor perfect set by itself and, consequently, homeomorphic to a Cantor set. The same topological property holds for each plane parallel to a coordinate plane which passes through the center of any of the chosen cubes of any rank. Therefore, each point of the universal Menger set has an arbitrarily small "cubic" neighbourhood whose boundary is homeomorphic to a Cantor perfect set. It follows that the universal Menger curve is one-dimensional and, hence, a curve in the sense of Uryson.

§ 2. The Definitions and Simplest Properties of the Dimensions ind, Ind, dim

2.1. Partitions

Definition 1. A closed subset C of a space X is called a partition between two disjoint subsets A_1 and A_2 if its complement $X \backslash C$ is a disjoint union of open sets

U_1 and U_2 which are neighbourhoods of A_1 and A_2, respectively. In this case, C is said to *separate* A_1 and A_2.

We remark that the sets U_1 and U_2 are not uniquely determined. For example, in the three point discrete space $X = \{x_1, x_2, x_3\}$, the empty set separates the points x_1 and x_2 and we can take U_1 to be either the one-point set $\{x_1\}$ or the two-point set $\{x_1, x_3\}$.

A partition is called *thin* if its interior Int C with respect to X is empty. Each partition contains thin partition. Indeed, suppose C separates A_1 and A_2 and $X\backslash C = U_1 \cup U_2$ where $U_i \supset A_i$. Put $V_1 = U_1, V_2 = X\backslash \bar{U}_1, D = \text{Bd } U_1 \backslash U_1$. Then the disjoint union $X = V_1 \cup D \cup V_2$ determines a thin partition D for A_1 and A_2. As a rule, this remark allows us to restrict our attention to thin partitions in what follows.

2.2. The Inductive Dimensions Ind and ind

Definition. Set Ind $X = -1$ if and only if the space X is empty. Suppose that the class of spaces X for which Ind $X \leqslant n - 1$, where n is a nonnegative integer, has already been defined. For each nonempty space X, set Ind $X \leqslant n$ if any two disjoint closed subsets A_1 and A_2 of X can be separated by a set C for which Ind $C \leqslant n - 1$.

If the condition Ind $X \leqslant n$ holds and the condition Ind $X \leqslant n - 1$ does not, then we will say that Ind $X = n$. If the condition Ind $X \leqslant n$ is not satisfied for any natural number n, then we shall say that Ind $X = \infty$ where we write $n < \infty$ for any $n = -1, 0, 1, 2, \ldots$.

The topological invariant Ind defined in this way is called the *large inductive dimension* (of the topological space X).

The definition of the *small inductive dimension* ind X is similar to the definition of Ind X except that one considers partitions for the pairs {point, closed set not containing the point} instead of all pairs of closed disjoint sets.

If a given point $x \in X$ and any closed set $A \subset X\backslash\{x\}$ can be separated by a set C of dimension ind $C \leqslant n - 1$, we write $\text{ind}_x X \leqslant n$. Here, the equality $\text{ind}_x X = n$ means that $\text{ind}_x X \leqslant n$ and the condition $\text{ind}_x X \leqslant n - 1$ does not hold. The invariant $\text{ind}_x X$ defined in this way is called the *small inductive dimension of the space X at the point x*. From the definitions we immediately obtain the equality

$$\text{ind } X = \sup\{\text{ind}_x X : x \in X\} \tag{3}$$

It follows from the remarks about thin partitions in §2.1 that the dimensions Ind and ind could have been defined as follows:

set Ind $X = -1$ and ind $X = -1$ if and only if $X = \varnothing$;

set Ind $X \leqslant n$ if, for any closed set $A \subset X$ and any neighbourhod OA of A, there is a neighbourhood $O_1 A$ such that $\overline{O_1 A} \subset OA$ and Ind Bd $O_1 A \leqslant n - 1$.

Similarly, set $\text{ind}_x X \leqslant n$ if for each neighbourhood Ox there is a neighbourhood $O_1 x$ such that $\overline{O_1 x} \subset Ox$ and ind Bd $O_1 x \leqslant n - 1$. The invariant ind X is defined by equality (3).

In particular, ind $X = 0$ if $X \neq \emptyset$ and the space has a base of open-closed sets. Similarly, Ind $X \leqslant 0$ means that any neighbourhood OA of any closed subset $A \subset X$ contains an open-closed neighbourhood.

2.3. The Simplest Properties of the Dimensions Ind and ind. The following properties are immediate from the definition:
1) if Ind X is finite, then the space X is normal;
2) if ind X is finite, the space X is regular;
3) if ind $X = 0$, the space X is completely regular.

In the latter case, points and disjoint closed sets are functionally separated by characteristic functions of open-closed sets.

As was mentioned in the introduction, all the spaces we consider are Hausdorff and, hence, satisfy the separation axiom T_1. Thus, a simple induction establishes that

$$\text{ind } X \leqslant \text{Ind } X. \tag{4}$$

It is obvious that if C separates A_1 and A_2 in X and $X_0 \subset X$, then $C \cap X_0$ separates $A_1 \cap X_0$ and $A_2 \cap X_0$ in X_0. Together with an induction on the dimension, this proves the following propositions.

Proposition 1 (*Monotonicity of the dimension* ind). *If* $X_0 \subset X$, *then* ind $X_0 \leqslant$ ind X.

Proposition 2 (*Monotonicity of* Ind *with respect to closed subsets*). *If* X_0 *is closed in* X, *then* Ind $X_0 \leqslant$ Ind X.

The following facts are left as easy exercises for the reader:
4) ind $\mathbb{R} = 1$;
5) if $X \subset \mathbb{R}$, then ind $X \leqslant 0$ if and only if $\mathbb{R}\backslash X$ is everywhere dense in $\mathbb{R}$;
6) ind $X = $ Ind X for each set $X \subset \mathbb{R}$;
7) if X is completely regular space and has cardinality less than c, then ind $X = 0$;
8) ind $\mathbb{R}^n \leqslant$ ind $S^n \leqslant n$.

The last assertion is proved by induction of n.

2.4. Other Inductive Dimensional Invariants. Doubtless the reader has already noted the asymmetry between the definition of the small inductive dimension ind by means of partitions and the definition of the large inductive dimension. This asymmetry can be eliminated as follows.

We define a dimensional *invariant* ind^p as follows. Set $\text{ind}^p X = -1$ if and only if $X = \emptyset$. If $X \neq \emptyset$, set $\text{ind}^p X \leqslant n$ if and only if any two distinct points x_1, $x_2 \in X$ can be separated by a set C of dimension $\text{ind}^p C \leqslant n - 1$.

We have the obvious inequality

$$\text{ind}^p X \leqslant \text{ind } X. \tag{5}$$

Proposition 3. *If* X *is a locally compact metrizable space, then*

$$\text{ind}^p X = \text{ind } X. \tag{6}$$

We give the proof in the case when $\operatorname{ind}^p X = 0$ and X is compact. Let $x \in X$ and Ox be a neighbourhood of x. For any point $y \in X \backslash Ox$, the space X is a disjoint union of open-closed subsets $U_y^1 \ni x$ and $U_y^2 \ni y$. From the cover $\{U_y^2; y \in X \backslash Ox\}$ of the compactum $X \backslash Ox$, we choose a finite subcovering $U_{y_1}^2, \ldots, U_{y_s}^2$. Then the set $U = \bigcap \{U_{y_i}^2: i = 1, \ldots, s\}$ will be an open-closed neighbourhood of x contained in Ox. The metrizability of the space X is used later: to establish the inequality

$$\operatorname{ind}^p X \geqslant \operatorname{ind} X$$

in positive dimensions, one has to apply the finite sum theorem for the dimension ind.

Even in the case of separable metric spaces, local compactness is necessary for equality (6) to hold.

Example. Let l_2^r be the subspace of the Hilbert space l_2 consisting of those $(t_i) \in l_2$ for which all coordinates are rational. It is clear that $\operatorname{ind}^p l_2^r = 0$. At the same time

$$\operatorname{ind} l_2^r \geqslant 1. \tag{7}$$

To verify inequality (7), it suffices to show that l_2^r has no nonempty bounded open-closed subsets. For this, in turn, it suffices to show that each neighbourhood U of $(0, 0, \ldots) \in l_2^r$ of radius less than or equal to 1 has boundary points. We inductively define a sequence of rational numbers $t_1, t_2, \ldots$ such that $x_m = (t_1, t_2, \ldots, t_m, 0, 0, \ldots) \in U$ and $\rho(x_m, l_2^r \backslash U) \leqslant 1/m$ where $m = 1, 2, \ldots$. For $m = 1$, we can set $t_1 = 0$. To pass from m to $m + 1$ it suffices to note that the set $A_{m+1} = \{(t_1, t_2, \ldots, t_m, t, 0, \ldots): t \text{ rational}\} \subset l_2^r$ is isometric to the space of rational numbers on the line and intersects both U and the complement of U. The point $x = (t_1, t_2, \ldots, t_m, \ldots)$ is a boundary point of the set U.

The first strictly inductive definition of dimension was given by Brouwer. In 1913, using an idea sketched the preceding year by Poincaré, he defined the *dimension* Dg (Dimensiongrad) and proved that

$$\operatorname{Dg} \mathbb{R}^n = n. \tag{8}$$

For a nonempty space X, we set $\operatorname{Dg} X = 0$ if and only if all subcontinua of X consist of a single point; that is, if and only if X is totally disconnected (see §3). Furthermore, we set $\operatorname{Dg} X \leqslant n$ if, for any two closed, disjoint subsets A_1, A_2, there exists a closed set $C \subset X \backslash A_1 \cup A_2$ of dimension $\operatorname{Dg} C \leqslant n - 1$ which *cuts* the space between A_1 and A_2. The latter means that every continuum K joining the set A_1 and A_2 necessarily intersects C.

Theorem 2. *If X is a locally connected, complete metric space, then*

$$\operatorname{Dg} X = \operatorname{Ind} X \tag{9}$$

2.5. The Definition of the Dimension dim. Before giving the definition of the *Lebesgue dimension* dim, we formulate an assertion which, like the lemma about shrinking covers (see article I of this volume, §2), characterizes normal spaces.

The Combinatorial Thickening Lemma. *For every finite collection* $\varphi = \{F_1, \ldots,$ $F_k\}$ *of closed subsets of a normal topological space* X *there exists a similar collection* $u = \{U_1, \ldots, U_k\}$ *such that* U_i *is a neighbourhood of* F_i. *Here similar means that,*

$$U_{i_1} \cap \cdots \cap U_{i_m} \neq \varnothing \Rightarrow F_{i_1} \cap \cdots \cap F_{i_m} \neq \varnothing.$$

It is also possible to assume the collection $\bar{u} = \{\bar{U}_1, \ldots, \bar{U}_k\}$ *is similar to* φ.

The Fundamental Definition. For a space X, we set dim $X \leqslant n$ if every finite open cover u of X has a refinement by a finite open cover v of multiplicity less than or equal to $n + 1$. If X satisfies the inequality dim $X \leqslant n$, but not dim $X \leqslant n - 1$, then we write dim $X = n$. Finally if the inequality dim $X \leqslant n$ does not hold for any n, we say that X is *infinite dimensional* and write dim $X = \infty$.

Definition 2. Let v be a cover of a space X which is a refinement of another cover u. Suppose the elements of u are indexed by ordinal numbers: $u = \{U_\alpha:$ $\alpha < \beta\}$. For $\alpha < \beta$ we put $W_\alpha = \bigcup \{V \in v: V \subset U_\alpha$ and $V \not\subset U_\gamma$ for $\gamma < \alpha\}$. The collection $w = \{W_\alpha: \alpha < \beta\}$ will be called an *enlargement of the cover* v with respect to the cover u and denoted $u(v)$.

It is easy to verify that $u(v)$ is a cover of X which combinatorially refines u and which has multiplicity $\text{ord}_x u(v)$ less than or equal to the multiplicity $\text{ord}_x v$ of the cover v at each point $x \in X$.

The following assertion gives a number of equivalent definitions of the dimension dim.

Proposition 4. *The following properties of a normal space* X *are equivalent*:
  (i) dim $X \leqslant n$;
every finite open cover of X *has*
  (ii) *an open refinement which is a cover of multiplicity* $\leqslant n + 1$;
  (iii) *an open combinatorial refinement which is a cover of multiplicity* $\leqslant n + 1$;
  (iv) *a closed combinatorial refinement which is a cover of multiplicity* $\leqslant n + 1$;
  (v) *a finite closed refinement which is a cover of multiplicity* $\leqslant n + 1$.

One proves that (i) $\Rightarrow$ (ii) $\Rightarrow$ (iii) $\Rightarrow$ (iv) $\Rightarrow$ (v) $\Rightarrow$ (i). Showing (ii) $\Rightarrow$ (iii) proceeds by enlarging covers; (iii) $\Rightarrow$ (iv) by shrinking covers, and (v) $\Rightarrow$ (i) by thickening covers.

Proposition 5. *If* X *is a metric compactum, then* dim $X \leqslant n$ *if and only if, for each* $\varepsilon > 0$, *there exists a closed finite* ε-*cover of multiplicity less than or equal to* $n + 1$.

Necessity follows from condition (v) of Proposition 4. Sufficiency follows from the existence of the Lebesgue number of an open cover of a metric compactum and condition (v) again.

Remark 1. It follows from Proposition 4 that we need not require that the refinement v be finite in the definition of dim. However, it is impossible to drop the finiteness requirement on the cover u. It is intuitively clear that the linearly ordered space $W(\omega_1)$ of transfinite numbers less than ω_1 must have dimension 0. But any open cover v which refines the cover $u = \{U_\alpha : \alpha < \omega_1\}$ where $U_\alpha = \{\beta : \beta < \alpha\}$ has multiplicity ω_1. We will prove below (chap 3, § 4) that the finiteness condition on u in the definition of dim can be weakened to a local finiteness condition.

It what follows, the dimension of a space X shall always mean the dimension dim X; when we have some other inductive dimension in mind, we will explicitly say so.

2.6. The Simplest Properties of the Dimension dim

Proposition 6 (Monotonicity of the dimension dim with respect to closed sets). *If X_0 is closed in X, then* dim $X_0 \leqslant$ dim X.

Proof. If $u = \{U_1, \ldots, U_s\}$ is an open cover of X_0 then $w = \{W_1, \ldots, W_s\}$, where $W_i = U_i \cup (X \backslash X_0)$, is an open cover of X. It has a refinement by an open cover v of multiplicity less than or equal to $n + 1$; the trace of the latter in X_0 is the desired refinement of u.

Theorem 3. *For any space X*

$$\dim X = 0 \Leftrightarrow \text{Ind } X = 0. \tag{10}$$

Proof. The implication $\Rightarrow$: Let A be a closed subset of X and OA any neighbourhood of A. Choose a cover $v = \{V_1, V_2\}$ of multiplicity 1 which combinatorially refines the cover $u = \{OA; X \backslash A\}$. Then V_1 will be an open-closed neighbourhood of A in OA.

The implication $\Leftarrow$: According to the shrinking lemma, the open covering $u = \{U_1, \ldots, U_s\}$ has a closed combinatorial refinement $\varphi = \{F_1, \ldots, F_s\}$. Since Ind $X = 0$, there exists an open-closed set W_i such that $F_i \subset W_i \subset U_i$, $i = 1, \ldots,$ s. We put $V_i = W_i \backslash \bigcup \{W_j : j < i\}$. Then $v = \{V_1, \ldots, V_s\}$ is a disjoint cover which is a combinatorial refinement of u.

§ 3. Zero-Dimensional Spaces

3.1. The Cantor Perfect Set. Let C be the subset of the segment $I = [0, 1]$ in the number line consisting of all numbers whose ternary representation does not contain the number 1; that is, the set of all numbers of the form $x = \sum_{i=1}^{\infty} 2x_i/3^i$ where x_i is equal to 0 or 1 for $i = 1, 2, \ldots$. We have $C = \bigcap_{i=1}^{\infty} C_i$ where C_i is the subset of I consisting of all numbers whose ternary representation does not have a 1 in the first i places. The sets C_i can be described somewhat differently as follows. Let C_1 be the subset obtained by dividing I into three equal subintervals and deleting the middle interval $(1/3, 2/3)$; that is, C_1 is the disjoint union of two intervals of length $1/3$. Let C_2 be the set obtained from C_1 by dividing each of the remaining intervals into three equal parts and deleting the middle parts; that is, $C_2 = C_1 \backslash \{(1/9, 2/9) \cup (7/9, 8/9)\}$. Thus C_2 is the disjoint union of four intervals of length $1/9$. We continue inductively: C_n is a disjoint union of 2^n intervals Δ_k of length $1/3^n$. We divide each interval Δ_k into three equal subintervals and throw away the middle ones to obtain the set C_{n+1}.

The set C defined in this way is called the *Cantor perfect set* or the *Cantor discontinuum*. The complement $\mathbb{R}\backslash C$ is everywhere dense in $\mathbb{R}$; hence, ind $C = 0$ (see Property 5 of Section 2.3). It is easy to verify that C has no isolated points and is, therefore, a *perfect set*.

The Cantor perfect set is homeomorphic to the product D^ω of countably many copies of the simple two point space.

Indeed, the map h which associates the sequence $(x_1, x_2, \ldots) \in D^\omega$ to the point $x = \sum_{i=1}^\infty 2x_i/3^i$ of the Cantor discontinuum is a bijection between C and D^ω. Furthermore, let $V = V_{j_1 \ldots j_n}$ be the basic set in D^ω defined by fixing the choice $\{j_1, \ldots, j_n\}$ of the first n coordinates. Then $h^{-1}V = \Delta_{j_1 \ldots j_n} \cap C$ where $\Delta_{j_1, \ldots, j_n}$ is the set of components of C_n (that is, intervals of length $1/3^n$) consisting of the points $x = \sum_{i=1}^\infty 2x_i/3^i$ satisfying $x_k = j_k$ for $1 \leqslant k \leqslant n$. Thus, the map h is continuous and, being a one to one and onto map (that is, a *condensation*) of a compactum to a Hausdorff space, is a homeomorphism.

Since C and D^ω are homeomorphic, any finite, or even countable, power of the Cantor perfect set is homeomorphic to itself. In particular, the square $C \times C$ is homeomorphic to C.

There is a purely topological characterization of the Cantor perfect set.

Theorem 4. *Up to homeomorphism, the Cantor perfect set is the only zero-dimensional metrizable compactum which does not have isolated points.*

Let X be a zero-dimensional metrizable compact space without isolated points. Fix a metric of X. The proof reduces to constructing a sequence $\{u_n\}$ of disjoint covers by open-closed sets on X such that:

a) u_n consists of 2^n elements $U_{i_1 \ldots i_n}$ where $i_k = 0, 1$;

b) $U_{i_1 \ldots i_n} = U_{i_1 \ldots i_n 0} \cup U_{i_1 \ldots i_n 1}$;

c) diam $U_{i_1 \ldots i_n} \to 0$ as $n \to \infty$.

We leave it to the reader to show that one can satisfy property c).

Let f be the map which takes the point $x = \sum_{i=1}^\infty 2x_i/3^i$ of the Cantor perfect set to the point $f(x) = \sum_{i=1}^\infty x_i/2^i$ of the interval I. The map $f\colon C \to I$ is continuous and onto. It has multiplicity two and glues together the endpoints of the intervals which are deleted from the closed segment in constructing the Cantor perfect set.

We can use f to construct an example of a Peano curve. Let $g\colon C \to C \times C$ be the homeomorphism whose existence was mentioned above. Then the epimorphism $\varphi = (f \times f) \circ g\colon C \to I^2$ extends to a map $\bar\varphi\colon I \to I^2$ as follows: if t_1 and t_2 are the endpoints of an interval deleted from the closed interval in the construction of C, we demand that $\bar\varphi$ map the interval $[t_1, t_2] \subset I$ affinely to the interval $[\varphi(t_1), \varphi(t_2)] \subset I^2$.

If we extend the map f to a map $c\colon I \to I$ by putting the value of c on each deleted interval equal to the value of f on the endpoints, we obtain a continuous function whose graph is the *Cantor staircase* (see Fig. 6). The map c is an example of a continuous function which is not absolutely continuous. It is differentiable on a set of total measure and its derivative is equal to zero.

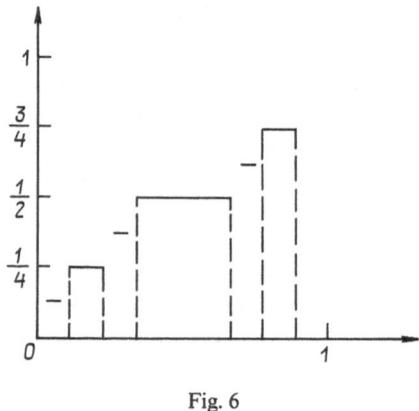

Fig. 6

3.2. Zero-Dimensionality in Comapct and Separable Metric Spaces. The Knaster-Kuratowski Fan.

Proposition 7. *If* ind $X = 0$ *and* X *is a Lindelöf space, then* dim $X = 0$.

Proof. Let $u = \{U_1, \ldots, U_s\}$ be an open cover of X. There is a countable refinement $w = \{W_1, \ldots, W_k, \ldots\}$ of u by open-closed sets. Put $V_k = W_k \backslash \bigcup \{V_i: i < k\}$. Then the family $v = \{V_1, \ldots, V_k, \ldots\}$ is a disjoint open cover which combinatorially refines w and is, therefore, a refinement of u.

A space X is called *totally discontinuous* if all subcontinua of X are single points: equivalently, if Dg $X \leqslant 0$ (see §2). A space X is called *totally disconnected* if every subspace containing more than one point is disconnected.

Theorem 5. *For a nonempty compactum X, the following are equivalent*:
  (i) X *is totally discontinuous*;
 (ii) X *is totally disconnected*;
(iii) $\text{ind}^p X = 0$;
 (iv) $\text{ind } X = 0$;
  (v) $\text{Ind } X = 0$;
 (vi) $\dim X = 0$.

Proof. The implication (i) $\Rightarrow$ (ii) is obvious and (iii) $\Rightarrow$ (iv) was established in the proof of Proposition 3. The equivalence of conditions (iv), (v) and (vi) follows from Theorem 3 and Proposition 7. The implication (iii) $\Rightarrow$ (i) is obvious. We need only verify the implication (ii) $\Rightarrow$ (iii). Let $x \in X$ and let Q_x denote the intersection of all open-closed subsets of X containing x. Suppose that Q_x is not connected. Then there exist nonempty closed disjoint sets A and B whose union is Q_x. Let $x \in A$. There exist disjoint neighbourhoods OA and OB in X. The collection of open-closed sets containing x is directed by inclusion. Therefore the neighbourhood $OA \cup OB$ of Q_x contains a neighbourhood U. Then $U \cap OA$ is an open-closed neighbourhood of x. Consequently, $Q_x \cap B = \emptyset$. This contradiction establishes that Q_x is connected. Since X is totally discontinuous, it follows that Q_x consists of a single point x and, hence, that $\text{ind}^p X = 0$.

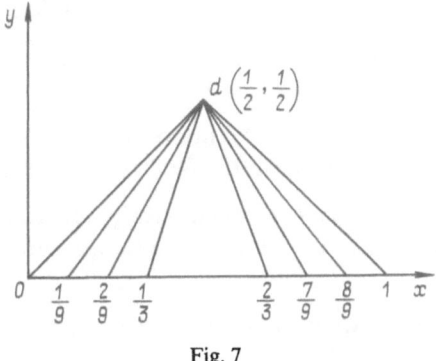

Fig. 7

For metric spaces with a countable base, it follows from Theorem 3 and Proposition 7 that conditions (iv), (v), and (vi) are equivalent. The implications (iv) $\Rightarrow$ (iii) $\Rightarrow$ (i) $\Rightarrow$ (ii) hold for any space X, but none of the converses hold in the class of separable metric spaces.

Indeed, the space l_2^r in Section 2.4 shows that the classes defined by conditions (iii) and (iv) are distinct. We now construct a space called the Knaster-Kuratowski fan and show that the classes defined by (i), (ii) and (iii) are distinct.

Consider the plane with a rectangular coordinate system and let C be the Cantor perfect set on the segment $[0, 1]$ of the axis of the abcissa. Let $d = (\frac{1}{2}, \frac{1}{2})$ and consider all straight line segments $[d, c]$ where c runs over all points of the set C. The union of these intervals will be denoted by D and is called the cone over the set C with vertex d (see Fig. 7)

We called $[d, c]$ a segment of the first type if the endpoint c is a point of the first type of the Cantor set C; that is, an endpoint of a deleted interval. The remaining intervals $[d, c]$ will be called intervals of the second type.

If $[d, c]$ is a segment of the first type, let D_c denote the subset consisting of all points (x, y) in $[d, c]$ whose ordinate is rational; if $[d, c]$ is an interval of the second type, let $D_c = \{(x, y) \in [d, c]: y \text{ is irrational}\}$. The union

$$K = \bigcup \{D_c : c \in C\}$$

is called the *Knaster-Kuratowski fan*. Set $K_0 = K \backslash d$.

It follows from the definition of K, that K is totally discontinuous. However, it is connected (see [I, Chap. 2, §4]). Therefore the Knaster-Kuratowski fan distinguishes between the classes (i) and (ii). Moreover, the space K_0 is totally disconnected. At the same time, two points a and b in the same D_c cannot be separated by the empty set. Indeed, if this were the case, it would easily follow that the set $K' = K \cap D'$, where $D' \subset C$ is the cone over some segment C' of the Cantor set C containing c, is disconnected. But K' is homeomorphic to K. Thus, the set K_0 distinguishes the classes (i) and (iii).

3.3. Some General Properties of Zero-Dimensional Sets

Proposition 8. *A product* $\prod \{X_\alpha \colon \alpha \in A\}$ *of inductively zero-dimensional spaces* X_α *(that is,* ind $X_\alpha = 0$*) is inductively zero-dimensional.*

This follows from the fact that a Tikhonov base of the product constructed using the bases of open-closed sets in the factors consists of open-closed sets.

Theorem 6. *A space* X *of weight* τ *is inductively zero-dimensional if and only if it is homeomorphic to a nonempty subset of the Cantor discontinuum* $D\tau$ *of weight* τ.

Proof. By Proposition 8, we have ind $D^\tau = 0$ and, according to Proposition 1, ind $X \leqslant 0$ for each $X \subset D^\tau$. Conversely, suppose ind $X = 0$ and $wX = \tau$. Then, there is a base $\mathscr{B} = \{B_\alpha \colon \alpha < \tau\}$ of open-closed sets in X of cardinality τ. Let φ_α be the characteristic function of the set B_α. Then the collection $\{\varphi_\alpha \colon \alpha < \tau\}$ functionally separates points and closed subsets of X. Therefore the Tikhnov map $\Delta\{\varphi_\alpha \colon \alpha < \tau\} \colon X \to D^\tau$ is an inclusion.

Corollary. *Zero-dimensional spaces with a countable base, and only these, are homeomorphic to subsets of the Cantor perfect set.*

Propositon 9 (The sum theorem for zero-dimensional sets). *Let* X *be a normal space which is a union of closed subsets* X_k *of dimension* Ind $X_k = 0$, $k = 1, 2, \dots$. *Then* Ind $X = 0$.

Proof. Let P and Q be disjoint closed subsets of X. There exist disjoint open-closed subsets A_1 and B_1 of X_1 such that

$$X_1 = A_1 \cup B_1, \quad P \cap X_1 \subset A_1, \quad Q \cap X_1 \subset B_1.$$

The sets $P \cup A_1$ and $Q \cup B_1$ are closed in X and disjoint. Therefore, there exist neighbourhoods G_1 and H_1, respectively, with disjoint closures in X. Replacing the sets P and Q by $\bar{G}_1$ and $\bar{H}_1$ in the preceding argument, we construct open sets G_2 and H_2 such that

$$G_2 \cup H_2 \supset X_2, \quad \bar{G}_2 \cap \bar{H}_2 = \varnothing, \quad \bar{G}_1 \subset G_2, \quad \bar{H}_1 \subset H_2.$$

Continuing this process, we obtain disjoint open sets $G = \bigcup \{G^k, k = 1, 2, \dots\}$ and $H = \bigcup \{H^k \colon k = 1, 2, \dots\}$ whose union is X and which contain P and Q respectively.

§4. The Addition Theorem for the Dimensions ind and dim

In this chapter we shall prove the inequalities

$$\text{ind}\,(M \cup N) \leqslant \text{ind}\,M + \text{ind}\,N + 1 \tag{11}$$

and

$$\dim(M \cup N) \leqslant \dim M + \dim N + 1 \tag{12}$$

for any sets M and N in a hereditarily normal space. These inequalities are called the Uryson-Menger formulas, having been established by Uryson and Menger for spaces with a countable base.

Immediate generalizations are the inequalities

$$\text{ind}(M_0 \cup M_1 \cup \cdots \cup M_n) \leqslant \text{ind } M_0 + \cdots + \text{ind } M_n + n \qquad (13)$$

and

$$\dim(M_0 \cup M_1 \cup \cdots \cup M_n) \leqslant \dim M_0 + \cdots + \dim M_n + n \qquad (14)$$

In particular, we have the following result.

Proposition 10. *If a hereditarily normal space X is a union of $n + 1$ inductively zero-dimensional sets, then* ind $X \leqslant n$.

Proposition 11. *If a hereditarily normal space X is a union of $n + 1$ sets M_i of dimension* dim $M_i \leqslant 0$, *then* dim $X \leqslant n$.

Before verifying inequalities (11) and (12), we remark that the following assertion holds and, in fact, characterizes hereditarily normal spaces.

Čech's Lemma. *Let $\gamma = \{\Gamma_1, \ldots, \Gamma_s\}$ be a collection of open sets of a subspace M of a hereditarily normal space X. Then there exist open subsets G_i of X such that $\Gamma_i = M \cap G_i$ and the collection $g = \{G_1, \ldots, G_s\}$ is similar to γ.*

A complete proof, which can be found in [AP], proceeds by induction on s. We consider the special case $s = 2$. Suppose $\Gamma_1 \cap \Gamma_2 = \varnothing$. There exists an open subset H_i of M such that $\Gamma_i = M \cap H_i$. Set $Y = H_1 \cup H_2$ and $A_i = \bar{\Gamma}_i^Y$. Then $A_1 \subset \bar{\Gamma}_1 \subset X \setminus H_2$ and $A_2 \subset \bar{\Gamma}_2 \subset X \setminus H_1$. Therefore, $A_1 \cap A_2 \subset X \setminus H_1 \cup H_2$ and, hence, $A_1 \cap A_2 = \varnothing$. Thus A_1 and A_2 are disjoint closed subsets of the normal space Y which is open in X. They have disjoint neighbourhoods G_1 and G_2 and these are the desired sets.

Inequality (11) is proved by induction on $m + n$ where $m = $ ind M and $n = $ ind N. For $m + n = -2$ it is obvious. Let $x \in M$, let O_x be any neighborhood of x in the space $X = M \cup N$, and let O_{1x} be a neighborhood of x whose closure lies in O_x. There exists a neighborhood Γ_1 of x in M such that $\Gamma_1 \subset O_{1x} \cap M$ and $\text{ind}(\text{Bd}_M \Gamma_1) \leqslant m - 1$. Set $\Gamma_2 = M \setminus \bar{\Gamma}_1^M$. According to Čech's lemma, there exist disjoint open sets G_i, $i = 1, 2$, in X cutting out the sets Γ_i on M. We may assume that $G_1 \subset O_{1x}$. Then Bd $G_1 \subset X \setminus G_1 \cup G_2 \subset N \cup \text{Bd}_M \Gamma_1$ and, by the induction hypothesis, ind Bd $G_1 \leqslant$ ind $M + \text{ind}(\text{Bd}_M \Gamma_1) + 1 \leqslant n + (m - 1) + 1 = m + n$, from which it follows that $\text{ind}_x X \leqslant m + n + 1$.

We shall prove inequality (12). Let $X = M \cup N$, dim $M = m$, dim $N = n$, and suppose that $u = \{U_1, \ldots, U_s\}$ is an open cover of X. The cover $u|_M = \{U_1 \cap M,$ $\ldots, U_s \cap M\}$ of X can be combinatorially refined to an open cover $g = \{G_1, \ldots,$ $G_s\}$ of M of multiplicity less than or equal to $m + 1$. Applying Čech's lemma, we obtain a collection $v = \{V_1, \ldots, V_s\}$ of open subsets of X of multiplicity less than or equal to $m + 1$ where $V_i \cap M = G_i$. We may assume that $V_i \subset U_i$. Dealing similarly with the collection $u|_N$, we obtain a collection $w = \{W_1, \ldots, W_s\}$ of multiplicity less than or equal to $n + 1$. Then $v \cup w$ is a refinement of u by an open cover of multiplicity less than or equal to $(m + 1) + (n + 1) = (m + n + 1) + 1$; that is, dim $X \leqslant m + n + 1$.

Chapter 2
Dimension Theory of Spaces with a Countable Base

§ 1. Simplicial Complexes

1.1. Points in General Position in $\mathbb{R}^n$. Fix a rectangular coordinate system on $\mathbb{R}^n$ and identify points with their position vectors. The vector from a to b will be denoted either by $\overrightarrow{ab}$ or $b - a$. Adding a vector $x = \{x^1, \ldots, x^n\}$ to a point $a = (a^1, \ldots, a^n)$ gives a point $a + x = (a^1 + x^1, \ldots, a^n + x^n)$.

A plane of dimension k, $0 \leqslant k \leqslant n$, in $\mathbb{R}^n$ is the set of points b obtained by adding to some point a all possible vectors $x = \lambda_1 x_1 + \cdots + \lambda_k x_k$ which are linear combinations of k fixed linearly independent vectors $x_1, \ldots, x_k$.

The points $a_0, a_1, \ldots, a_k$ in $\mathbb{R}$ are called *geometrically* (or *affinely*) *independent* if the vectors $\overrightarrow{a_0 a_1}, \ldots, \overrightarrow{a_0 a_k}$ are linearly independent. This is equivalent to requiring that the points $a_0, a_1, \ldots, a_k$ lie in a (uniquely determined) k-dimensional plane $R(a_0, \ldots, a_k)$ and do not lie in any plane of dimension less than k.

Definition 1. A set of points in $\mathbb{R}^n$ is said to be in *general position* if every subset consisting of $n + 1$ or fewer points is geometrically independent.

A sequence $\bar{\varepsilon} = (\varepsilon_1, \varepsilon_2, \ldots)$ of positive numbers is called a *size*. Sizes are partially ordered coordinatewise. If $A = \{a_1, a_2, \ldots\}$ is a countable (ordered) subset of a metric space X and $\bar{\varepsilon}$ is a size, then we shall say that a map $f: A \to X$ is a *shift of size* less than $\bar{\varepsilon}$ if $\rho(a_i, f(a_i)) < \varepsilon_i$.

Proposition 1. *For any countable (or finite) set $A = \{a_1, a_2, \ldots\} \subset \mathbb{R}^n$ and any size $\bar{\varepsilon}$ there exists a shift $f: A \to \mathbb{R}^n$ of size less than $\bar{\varepsilon}$ carrying the set $A = \{a_1, a_2, \ldots\}$ to a set $B = \{f(a_1), f(a_2), \ldots\}$ in general position. Conversely, if a countable set $A = \{a_1, a_2, \ldots\} \subset \mathbb{R}^n$ is in general position, then there exists a size $\bar{\varepsilon}$ such that no translation of size less than $\bar{\varepsilon}$ takes the set out of general position.*

1.2. Simplices. Geometrical Complexes. Let $a_0, a_1, \ldots, a_k$ be a set of geometrically independent points in $\mathbb{R}^n$. Then the plane $R(a_0, \ldots, a_k)$ they determine consists of all points of the form

$$b = a_0 + \mu_1(a_1 - a_0) + \cdots + \mu_k(a_k - a_0)$$

or, what is the same,

$$b = \mu_0 a_0 + \mu_1 a_1 + \ldots \mu_k a_k$$

where

$$\mu_0 + \mu_1 + \cdots + \mu_k = 1. \tag{1}$$

The numbers $\mu_0, \mu_1, \ldots, \mu_k$ satisfying condition (1) are uniquely determined by the point b and are all called its *barycentric coordinates* (in the barycentric coordinate system consisting of the points $a_0, a_1, \ldots, a_k$)

Definition 2. Let $a_0, a_1, \ldots, a_k$ be a collection of geometrically independent points in $\mathbb{R}^n$. The set of points $b \in \mathbb{R}^n$ all of whose barycentric coordinates in the system $a_0, \ldots, a_k$ are positive is called an *open k-dimensional simplex* with vertices $a_0, \ldots, a_k$ and is denoted by $T^k = |a_0, \ldots, a_k|$. The set of points all barycentric coordinates of which are nonnegative is called the *closed k-dimensional simplex* $\bar{T}^k = \overline{a_0 \ldots a_k}$ with vertices $a_0, \ldots, a_k$.

It is easy to verify the following assertions.
1) A closed simplex $\bar{T}^k$ is the closure of an open simplex T^k in $\mathbb{R}^n$ and an open simplex T^k is the interior of the closed simplex $\bar{T}^k$ relative to the subspace $R(a_0, \ldots, a_k)$ of R^n.
2) Both open and closed simplices with vertices $a_0, \ldots, a_k$ are convex sets in R^n and the closed simplex $\bar{T}^k$ is the convex hull of the set $\{a_0, \ldots, a_k\}$.
3) The set $\{a_0, \ldots, a_k\}$ of vertices of a simplex, called the *skeleton*, is the set of extreme points of the closed simplex $\bar{T}^k$ and is, therefore, uniquely determined by the simplex.

Definition 3. Let $a_{i_0}, \ldots, a_{i_r}$ be vertices of a k-dimensional simplex $T^k = |a_0, \ldots, a_k|$. These vertices are geometrically independent and define an open simplex T^r and a closed simplex $\bar{T}^r = \overline{a_{i_0} \ldots a_{i_r}}$ in the plane $R(a_{i_0}, \ldots, a_{i_r}) \subset R(a_0, \ldots, a_k)$, called *open* and *closed r-dimensional faces of the simplex* T^k, respectively (with vertices $a_{i_0}, \ldots, a_{i_r}$). The simplex T^k is itself a (improper) face. Let $\bar{T}_i^{k-1}$ denote the closed face of T^k opposite the vertex a_i.

The following assertions are elementary.
4) For $0 \leqslant r \leqslant h \leqslant k$ a closed face $\bar{T}^r$ is contained in a closed face $\bar{T}^h$ if and only if the skeleton of T^r is contained in the skeleton of T^h.
5) The intersection of any two faces of a simplex T^k is either empty or is a face.
6) The intersection of all *closed* $(k-1)$-dimensional faces of T^k is empty.
7) If ε is the Lebesgue number of the collection of all $(n-1)$-dimensional closed faces of a simplex T^k and if a set $M \subset \mathbb{R}^n$ of diameter less than ε contains a vertex $a_i \in \bar{T}^k$, then M does not intersect the face $\bar{T}_i^{k-1}$ opposite the vertex.

Unless explicitly stated otherwise, we use the word "simplex" below to refer to an open simplex.

A *geometrical simplicial complex* is a set of simplices (in a fixed space $\mathbb{R}^n$). A complex K is called *n-dimensional* if all its simplices have dimension less than or equal to n and at least one has dimension n. A complex K is called *complete* if it contains every face of every simplex in K.

Definition 4. A complete finite complex K whose elements are pairwise disjoint simplices in a given space $\mathbb{R}^n$ is called a *triangulation* lying in $\mathbb{R}^n$.

The *underlying polyhedron* (or *body*) of a complex K is the union of all simplices which are elements of K. It is denoted by $\tilde{K}$. If a complex K consists of a single simplex T, then the underlying polyhedron $\tilde{K} = \tilde{T}$ will be denoted simply by T.

A compactum which is the underlying polyhedron of a triangulation is called a *polyhedron*; if $P = \tilde{K}$ then the triangulation K is called a *triangulation of the polyhedron P*. Every non-zero-dimensional polyhedron has infinitely many triangulations.

A *subcomplex* of a complex K is any nonempty subset $K_0 \subset K$.

1.3. Abstract Simplicial Complexes. In Section 2, we observed that there is a one to one correspondence in $\mathbb{R}^n$ between simplices and their skeleta. In view of this, it is natural to introduce the following definition.

Definition 5. Let E be a set of elements called vertices. An *abstract simplicial complex* (on a given set E of vertices) is a collection K of finite subsets T of E called *abstract simplices* or *skeleta* of the complex K.

A *face* of an abstract simplex is a subset of it; the *dimension of an abstract simplex* is one less than the number of its vertices; the *dimension of a complex K* is defined to be the upper bound of the dimensions of the abstract simplices $T \in K$.

Examples. 1. A geometric simplex can clearly be considered to be an abstract simplex.

2. Let $\alpha = \{A_1, \ldots, A_s\}$ be a finite collection of sets. Each element A_i can be put into correspondence with some vertex a_i (most simply, although not always most conveniently, by taking a_i to be the set $A_i \in \alpha$). A set of vertices is a simplex if and only if the corresponding sets have nonempty intersection. The (complete) abstract complex obtained in this way is called the *nerve* of a collection α of sets and denoted by $N(\alpha)$.

1.4. Stars. Open and Closed Subcomplexes. The *combinatorial star of a simplex T in a complex K* is the set $O_K T$ of all simplices of K which have T as a face. The following can be verified immediately.

Proposition 2. *The combinatorial stars $O_K T_1, \ldots, O_K T_r$ of a complete simplicial complex have nonempty intersection if and only if the union of the simplices T_1, $\ldots$, T_r is a simplex $T \in K$ called the combinatorial union of the simplices $T_1, \ldots$, T_r. In this case*

$$O_K T_1 \cap O_K T_2 \cap \cdots \cap O_K T_r = O_K T.$$

A subcomplex K_0 of a complex K is called *open* in K if the combinatorial star of each simplex $T \in K_0$ in K_0 is contained in K_0. The complement of an open subcomplex is called *closed*.

Proposition 3. *In order that the underlying polyhedron $\tilde{K}_0$ of a subcomplex $K_0 \subset K$ of a triangulation K be open in $\tilde{K}$ it is necessary and sufficient that K_0 be open in K.*

If K is a triangulation, then the underlying polyhedron of the combinatorial star of a simplex $T \in K$ is called an *open star* and denoted OT; open stars of vertices are called *principal* stars. Open stars are open sets by Proposition 3. Proposition 2 implies the following.

Proposition 4. *Every triangulation is the nerve of the collection of its principal stars.*

Corollary. *The collection $\{Oa_i\}$ of principal stars of a triangulation K of a polyhedron P is an open cover of multiplicity $n + 1$ where n is the dimension of the triangulation K.*

1.5. Barycentric Subdivision

Definition 6. A triangulation K_1 is called a *subdivision* of a triangulation K if K_1 is a refinement of K and if $\tilde{K}_1 = \tilde{K}$.

The most important subdivision of a triangulation K is its *barycentric subdivision K'* which is defined as follows. The *center of a simplex* $T^n = |a_0 \dots a_n|$ is defined to be the geometric center of gravity; that is, the point a all of whose barycentric coordinates are the same:

$$a = (1/(n + 1))a_0 + \cdots + (1/(n + 1))a_n. \tag{2}$$

The vertices of K' are the centers of the simplices of K and a collection of vertices is deemed to be a simplex of K' if they can ordered

$$e_0, e_1, \dots, e_r$$

such that the corresponding simplices $T_0, T_1, \dots, T_r$ form a decreasing sequence

$$T_0 > T_1 > \cdots > T_r \tag{3}$$

where the inequality $T_{i+1} < T_i$ means that T_{i+1} is a proper face of the simplex T_i.

It follows from (3) that the simplex $|e_0 \dots e_r| \in K'$ lies in the simplex $T_0 \in K$ and, hence, that K' is a refinement of K. The proof that $\tilde{K}' = \tilde{K}$ is left to the reader. The barycentric subdivision of a two dimensional simplex is sketched in Fig. 8. It consists of 6 two-dimensional, 12 one-dimensional and 7 zero-dimensional simplices (vertices).

It is well known that the diameter of a simplex is equal to the length of its longest edge. From the definition (2) of the center a of a simplex, it easily follows that the largest of the distances $\rho(a, a_i), i = 0, 1, \dots, n$, does not exceed $(n/(n + 1))\delta$ where δ is the diameter of the simplex $T^n = |a_0 \dots a_n|$. Hence, the mesh d of an n-dimensional triangulation K (that is the largest of the diameters of its simplices) is related to the mesh d' of its barycentric subdivision K' by the inequality

$$d' \leqslant (n/(n + 1))d. \tag{4}$$

Thus, repeatedly carrying out the operation of barycentric subdivision on a given triangulation K allows one to obtain subdivisions with arbitrarily small mesh.

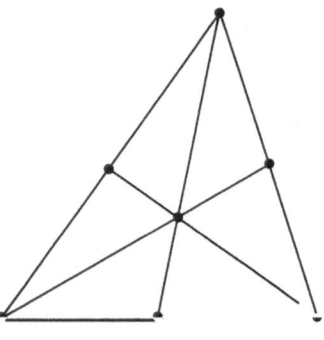

Fig. 8

1.6. Simplicial Maps. Suppose that to every vertex a of a simplical complex K is associated a vertex $b = f(a)$ of a simplicial complex L and that whenever a collection of vertices $a_{i_0} \ldots a_{i_r}$ generates a simplex in K, the vertices $f(a_{i_0})$, ..., $f(a_{i_r})$, some of which may coincide, generate a simplex in L. Then to each simplex T of K there corresponds a simplex $T' = f(T)$ of L and we obtain a map $f : K \to L$ called a *simplicial map* of the complex K to the complex L.

A one to one map of a complex K onto a complex L is called an *isomorphism* and the complexes K and L are said to be *isomorphic*.

Proposition 5. *Every finite n-dimensional complete simplicial complex K is isomorphic to a triangulation L lying in $\mathbb{R}^{2n+1}$.*

Proof. Let $a_1, \ldots, a_s$ be all the vertices of the complex K. According to Proposition 1, it is possible to take points $b_1, \ldots, b_s$ of $\mathbb{R}^{2n+1}$ in general position. We require that the points $b_{i_0}, \ldots, b_{i_r}, r \leqslant n$, span a simplex in $\mathbb{R}^{2n+1}$ if and only if $a_{i_0}, \ldots, a_{i_r}$ generate a simplex in K. The simplices in $\mathbb{R}^{2n+1}$ obtained in this way are pairwise disjoint. Indeed, let $T_1 = |b_{i_0} \ldots b_{i_h}|$ and $T_2 = |b_{j_0} \ldots b_{j_k}|$ be simplexes corresponding to different simplexes $|a_{i_0} \ldots a_{i_h}|$ and $|a_{j_0} \ldots a_{j_k}|$ of K. Then the union of the skeletons of these simplices consists of no more than $(h + 1) + (k + 1) \leqslant (n + 1) + (n + 1) = 2n + 2$ points and, therefore, being a set of points in general position in $\mathbb{R}^{2n+1}$, is the skeleton of some simplex T in $\mathbb{R}^{2n+1}$. The simplices T_1 and T_2 are open faces of T and therefore disjoint. The triangulation L in $\mathbb{R}^{2n+1}$ constructed in this way is isomorphic to the complex K and is called a *geometrical realization* of K.

Remark 1. If we do not require that the simplexes of L be disjoint, then it is clear from the proof of Proposition 5 that any n-dimensional simplicial complex can be geometrically realized in n-dimensional Euclidean space $\mathbb{R}^n$.

Let f be a simplicial map of a triangulation K to a triangulation L. Then the map of the vertices also defines an affine map of each simplex $T \in K$ to the simplex $f(T) \in L$. These affine maps are continuous on each simplex of the complex K and, taken together, form a continuous ("piecewise affine") map of the polyhedron $\tilde{K}$ to the polyhedron $\tilde{L}$ which is denoted by $\tilde{f}$ (and, sometimes, simply as f). The map $\tilde{f}$ is called a *simplicial map*.

§2. Sperner's Lemma and its Corollaries

2.1. Sperner's Lemma. *Let K be a triangulation of a closed simplex $\bar{T}^n = \overline{a_0 a_1 \ldots a_n}$. Suppose that each vertex e of K corresponds to a vertex $f(e)$ of the simplex $\bar{T}^n$ and that if a vertex $e \in K$ lies in a face $\bar{T}^r$ of $\bar{T}^n$, then $f(e) \in \bar{T}^r$. Then there exists at least one simplex τ^n of the triangulation K whose vertices correspond to different vertices of the simplex $\bar{T}^n$.*

The proof is by induction on n. One proves that the number of simplexes τ^n of K whose vertices correspond to different vertices of T^n is odd. We outline the first step of the induction ($n = 1$) which gives a reasonably good idea of how the proof goes in general.

Call a vertex e of the triangulation K distinguished if $f(e) = a_0$. A segment τ (a one-dimensional simplex) of K will be called:

a) nondegenerate if precisely one of its vertices is distinguished and

b) degenerate if both vertices are distinguished.

It is necessary to prove that the number α of all nondegenerate segments of K is odd.

Let β denote the number of degenerate segments. Consider the collection of all pairs (e, τ) of vertices e and segments τ of the triangulation K. A pair (e, τ) will be called distinguished if e is distinguished and a vertex of the segment τ. Since each nondegenerate segment occurs exactly once in a distinguished pair, and each degenerate vertex occurs in exactly two distinguished pairs, the number of all distinguished pairs must be $\alpha + 2\beta$. Let γ be the number of distinguished vertices of the triangulation K. Since each distinguished vertex lying in a segment $[a_0, a_1]$ belongs to two distinguished pairs and the vertex a_0 to one distinguished pair, we have $\alpha + 2\beta = 2\gamma - 1$ from which it follows that α is odd.

Proposition 6. *Let $\{A_0, A_1, \ldots, A_n\}$ be a closed cover of a closed simplex $\bar{T}^n$ such that each face $\bar{T}^r = \overline{a_{i_0} a_{i_1} \ldots a_{i_r}}$ is covered by the sets $A_{i_0}, A_{i_1}, \ldots, A_{i_r}$. Then $A_0 \cap A_1 \cap \cdots \cap A_n \neq \varnothing$.*

Proof. By inequality (4) of 1.5, there exist triangulations of any polyhedron with arbitrarily small mesh. Therefore, it is sufficient to prove that in any triangulation K of the closed simplex $\bar{T}^n$ there exists a simplex whose skeleton intersects all of the sets $A_0, A_1, \ldots, A_n$. To each vertex e of K we associate the vertex $f(e) = a_k \in \bar{T}^n$ such that $A_k \ni e$; here, if e lies on a face $\overline{a_{i_0} a_{i_1} \ldots a_{i_r}}$, then we associate to it the vertex of this face. Then the hypotheses of Sperner's lemma hold and the assertion follows.

2.2. The Equality dim $P^n = n$ for n-Dimensional Polyhedra. The inequality dim $P^n \leqslant n$ follows from Proposition 4 and the existence of triangulations of a polyhedron P^n of arbitrarily small mesh. To verify the opposite inequality it is sufficient to prove that dim $\bar{T}^n \geqslant n$.

Set $U_i = \bar{T}^n \backslash \bar{T}_i^{n-1}$. Then $u = \{U_0, \ldots, U_n\}$ is an open cover of the closed simplex $\bar{T}^n$. We shall show that every cover α which is a combinatorial refinement of u satisfies the hypotheses of Proposition 6. Suppose that $\kappa = \{i_0, i_1, \ldots, i_r\}$ and $\lambda = \{0, 1, \ldots, n\} \backslash \kappa$. Since $A_i \cap \bar{T}_i^{n-1} = \varnothing$, it follows that $(\bigcup \{A_i : i \in \lambda\}) \cap (\bigcap \{\bar{T}_i^{n-1} : i \in \lambda\}) = \varnothing$. But $\bigcap \{\bar{T}_i^{n-1} : i \in \lambda\} = \overline{a_{i_0} a_{i_1} \ldots a_{i_r}}$. Therefore, $\overline{a_{i_0} a_{i_1} \ldots a_{i_r}}$ lies in $\bigcup \{A_i : i \in \kappa\}$ which is what we were required to prove. Thus, we have shown that

$$\dim P^n = n. \tag{5}$$

Using this, we obtain the following proposition.

Proposition 7. *Every compact set X lying in an n-dimensional Euclidean space $\mathbb{R}^n$ has dimension* dim $X \leqslant n$.

Now we can prove *Brouwer's* important *theorem on the invariance of the dimension of Euclidean space.*

Theorem 1. *If $m \neq n$, the space $\mathbb{R}^m$ and $\mathbb{R}^n$ are not homeomorphic.*

This follows from the stronger assertion below.

Proposition 8. *For $m > n$ no set $E \subset \mathbb{R}^m$ containing an interior point can be topologically imbedded in $\mathbb{R}^n$.*

Indeed, E contains an m-dimensional simplex $\bar{T}^m$ and, by (5), dim $\bar{T}^m = m$. Therefore, the assertion that E is included in $\mathbb{R}^n$ contradicts Proposition 7.

2.3. Brouwer's Fixed Point Theorem and the Invariance of Interior Points of a Set $E \subset \mathbb{R}^n$

Theorem 2. *For every continuous map $f: \bar{T}^n \to \bar{T}^n$ there exists a point $x \in \bar{T}^n$ such that $f(x) = x$.*

Proof. Suppose that f carries a point x with barycentric coordinates $\mu_0, \ldots, \mu_n$ to a point $f(x)$ with barycentric coordinates $\mu_0', \ldots, \mu_n'$. We let A_i denote the set of all points $x \in \bar{T}^n$ for which $\mu_i' \leqslant \mu_i$. It is easy to check that $\{A_0, \ldots, A_n\}$ is a closed cover of the simplex $\bar{T}^n$ satisfying the hypotheses of Proposition 6. Then, for each point $x \in A_0 \cap \cdots \cap A_n$, the inequalities

$$0 \leqslant \mu_0' \leqslant \mu_0, \quad 0 \leqslant \mu_1' \leqslant \mu_1, \ldots, 0 \leqslant \mu_n' \leqslant \mu_n$$

hold. In view of the conditions

$$\mu_0' + \cdots + \mu_n' = \mu_0 + \cdots + \mu_n = 1$$

we have

$$\mu_0' = \mu_0, \mu_1' = \mu_1, \ldots, \mu_n' = \mu_n.$$

Thus, each point $x \in A_0 \cap \cdots \cap A_n$ is fixed.

Theorem 3. *A homeomorphism f of a set $E \subset \mathbb{R}^n$ to a set $f(E) \subset \mathbb{R}^n$ carries each interior point of E to an interior point of $f(E)$.*

It suffices to prove the theorem for $E = \overline{T}^n$. If $p \in \overline{T}^n$ is an interior point, then taking the cone over the barycentric subdivision of the boundary S^{n-1} of the simplex T^n with vertex at the point p gives a triangulation K of the closed simplex $\overline{T}^n$ differing from the barycentric subdivision only in that the "center" of the simplex T^n is taken to be the point p. Let A_i be the union of the closures of the simplexes of K which have the vertex $a_i \in \overline{T}^n$ as a vertex.

The point p is the unique common point of the sets $A_0, \ldots, A_n$. Every cover α' of $\overline{T}^n$ obtained from the cover $\alpha = \{A_0, \ldots, A_n\}$ by modifying the sets A_i in a sufficiently small neighbourhood of p (with closure lying in T^n) satisfies the hypotheses of Proposition 6 and, consequently, has multiplicity $n + 1$. Therefore, Theorem 3 will be proved if we can establish the following.

Lemma 1. *Let $\alpha = \{A_1, \ldots, {}_{as}\}$ be a closed cover of multiplicity $n + 1$ of a compact set $X \subset \mathbb{R}^n$ and suppose that $\operatorname{ord}_p \alpha = n + 1$ at only one point $p \in X$. If p is not an interior point of X, then it is possible to change the cover α in an arbitrarily small neighbourhood of the point p to obtain a cover α' of multiplicity less than or equal to n.*

Proof. Choose an arbitrarily small simplex T^n containing the point p. Each point of the boundary S^{n-1} of T^n belongs to no more than n elements of α. Since $\dim S^{n-1} \leqslant n - 1$ by (5), the collection $\{A_0 \cap S^{n-1}, \ldots, A_n \cap S^{n-1}\}$ can be thickened to a cover $\{B_0, \ldots, B_n\}$ of S^{n-1} of multiplicity less than or equal to n. Since p is not an interior point, there exists a point $q \in T^n$ not contained in X. Let C_i denote the cone over the set B_i with vertex at the point q and set

$$A'_i = (A_i \backslash T^n) \cup (C_i \cap X).$$

Then $\alpha' = \{A'_1, \ldots, A'_n\}$ is the desired modification of the cover α.

2.4. The Essentialness of the Identity Map of a Simplex

Definition 7. A continuous map $f: X \to \overline{T}^n$ is called *essential* if every continuous map $f_1: X \to \overline{T}^n$ which coincides with f on all points of the set $f^{-1}S^{n-1}$ is surjective (onto $\overline{T}^n$).

The following result is obvious.

Proposition 9. *A continuous map $f: X \to \overline{T}^n$ is inessential if and only if it can be swept onto the boundary; that is, if there exists a continuous map $f_1: X \to S^{n-1}$ coinciding with f on the set $f^{-1}S^{n-1}$.*

Theorem 4. *The identity map of a closed simplex to itself is essential.*

Proof. Suppose the converse. Replace the closed simplex $\overline{T}^n$ by the ball B^n homeomorphic to it. By Proposition 9, we obtain a continuous map $f: B^n \to S^{n-1}$ of the ball to its boundary which fixes all points of the boundary. Let $h: S^{n-1} \to S^{n-1}$ be the antipodal map. Then the map $g = h \circ f: B^n \to B^n$ does not have a fixed point, contradicting Theorem 2.

§3. The Approximation Theorems for ε- and ω-Mappings

3.1. ε-Shifts and ω-Mappings. We begin with the definitions of an ε-shifts, an ε-map, and an ω-map. Let X be a set which lies in a metric space M. A continuous map $f: X \to Y$ onto a set $Y \subset M$ will be called an ε-*shift* if, for a given positive ε, the inequality

$$\rho(x, f(x)) < \varepsilon$$

holds for each point $x \in X$.

A continuous map f of a metric space X into any space Y is called an ε-*map* if the pre-image $f^{-1}y$ of each point $y \in Y$ has diameter less than ε. Using the triangle inequality, it can be immediately verified that an ε-shift is a 3ε-map and even a $(2\varepsilon + \delta)$-map for any $\delta > 0$. It is easy to check that if $f: X \to Y$ is an ε-map of a compactum, then every point $y \in Y$ has a neighbourhood O_y for which

$$\text{diam } f^{-1}O_y < \varepsilon.$$

Thus, the following definition is the topological analogue of the definition of an ε-map for compacta.

Definition 8. Let ω be an open cover of the space X. Then a continuous map $f: X \to Y$ is called an ω-*map* if each point $y \in Y$ has a neighbourhood O_y whose pre-image $f^{-1}O_y$ is contained in some element of the cover ω.

It is easy to verify that, for metric compacta, the concepts of an ε-map and an ω-map are equivalent in the following sense.

Proposition 10. *A compactum X admits an ε-map, for each $\varepsilon > 0$, into (resp., onto) a compactum Y of a given class $\mathscr{K}$ if and only if, for each open cover ω of the compactum X, there exists an ω-map $f: X \to Y$ into (resp., onto) Y of the class $\mathscr{K}$.*

3.2. Barycentric and Canonical Maps. Let $\omega = \{O_1, \ldots, O_s\}$ be an open cover of a normal space X and $N = N(\omega)$ any geometric realization of the nerve of the cover. Let $\Phi(\omega) = \{\varphi_1, \ldots, \varphi_s\}$ be a partition of unity subordinate to the cover ω; that is, a collection of continuous functions such that

$$\sum_{i=1}^{s} \varphi_i = 1 \quad \text{and} \quad \varphi_i^{-1}(0, 1] \subset O_i.$$

We now construct a map f of the space X to the polyhedron $\tilde{N}$. First, define a map $f': X \to \mathbb{R}^s$ by the formula

$$f'(u) = (\varphi_1(x), \ldots, \varphi_s(x)). \tag{6}$$

The map f' is continuous and $f'(X)$ lies in the intersection of the nonnegative cone $\mathbb{R}^s_+$ with hyperplane $P \subset \mathbb{R}^s$ defined by the equation

$$t_1 + \cdots + t_s = 1,$$

where t_i is the ith coordinate of the point $t \in \mathbb{R}^s$. The intersection $\mathbb{R}_+^s \cap P$ is an $(s - 1)$-dimensional closed simplex $\bar{T}^{s-1} = \overline{a_1 \ldots a_s}$ where a_i is the unit point on the coordinate axis $\mathbb{R}_i \subset \mathbb{R}^s$. Here the functions φ_i are barycentric coordinates of the point $f'(x)$; or, rather, if $\mu_1, \ldots, \mu_s$ are barycentric coordinates of the point $f'(x)$, then $\varphi_i(x) = \mu_i$; that is, if the μ_i are considered as continuous functions on the simplex $\bar{T}^{s-1}$, then

$$\varphi_i = \mu_i \circ f'.$$

Now note that the polyhedron $\tilde{N}$ with given triangulation N is isomorphically imbedded in the simplex $\bar{T}^{s-1}$. Indeed, if e_i is a vertex of the nerve of N corresponding to the set O_i, then setting $h(e_i) = a_i$ gives a map h of the set of vertices of N to the skeleton of the simplex $\bar{T}^{s-1}$ which extends uniquely to a simplicial imbedding $h \colon \tilde{N} \to \bar{T}^{s-1}$. Thus, we also obtain another geometric realization of the complexes.

Proposition 11. *Each finite simplicial complex K is isomorphic to a subcomplex of an elementary simplicial complex $K(T^n)$ formed by all faces of the simplex $\bar{T}^n$.*

Now set $f = h^{-1} \circ f'$ to obtain a continuous map $f \colon X \to \tilde{N}$ called a *barycentric map* (with respect to the partition of unity $\Phi(\omega)$). Note that, as above, the numbers $\varphi_i(x)$ are barycentric coordinates of the point $f(x)$ in the barycentric coordinate system $\{e_1, \ldots, e_s\}$. Here, we have extended somewhat the concept of a barycentric coordinate system. Previously, we applied it only in the context of a single simplex; we now apply it to the set of all vertices of a complex. Here, as before,

$$p \in T^r = |e_{i_0} \ldots e_{i_r}| \Leftrightarrow \mu_{i_k}(p) > 0 \quad \text{for} \quad k = 0, \ldots, r.$$

In particular,

$$Oe_i = \{p \in \tilde{N} \colon \mu_i(p) > 0\}. \tag{7}$$

We call a map f of a space X to a polyhedron $\tilde{N}$, where, as above, $N = N(\omega)$ is the nerve of a cover ω, *canonical* (with respect to the cover ω) if the preimage $f^{-1}Oe_i$ of each principal star Oe_i is contained in O_i. It is obvious that every canonical map is an ω-map. According to (7), each barycentric map is canonical. Conversely, we have the following proposition.

Proposition 12. *Every canonical map $f \colon X \to \tilde{N}$ is barycentric with respect to a partition of unity $\{\mu_1 \circ f, \ldots, \mu_s \circ f\}$ where the μ_i, $i = 1, \ldots, s$, are functions of the barycentric coordinates in the complex N.*

We shall say that a map $f \colon X \to \tilde{K}$ essentially *covers* a simplex $T \in K$ if the map $f \colon f^{-1}\bar{T} \to \bar{T}$ is essential.

Theorem 5 (The canonical map theorem). *Let $\omega = \{O_1, \ldots, O_s\}$ be an open cover of a normal space X with nerve N realized as a triangulation. Then there exists a subcomplex N' of N and a canonical (with respect to ω) map $f \colon X \to \tilde{N}$ for which $f(X) = \tilde{N}'$ and each principal simplex of $\tilde{N}'$ is essentially covered.*

The theorem is proved by taking a canonical map $g: X \to \tilde{N}$ and modifying it by successively sweeping out principal simplexes which are covered inessentially. The process of sweeping out consists of the following. Let $T \in N$ be a principal simplex which is inessentially covered by the map g. This means that there exists a map $h: g^{-1}\bar{T} \to S$, where S is the boundary of T, such that $h = g$ on $g^{-1}S$. Set

$$g_1(x) = \begin{cases} g(x) & \text{if } x \in g^{-1}(\tilde{N} \setminus T); \\ h(x) & \text{if } x \in g^{-1}\bar{T}. \end{cases}$$

We say that the continuous map g_1 is obtained by *sweeping g out* of the inessentially covered simplex T.

Remark 2. One can also define a barycentric map $f: X \to \tilde{N}$ when the geometric realization of the nerve N of a cover ω is not a triangulation. To do this, we first take a barycentric map $f_1: X \to \tilde{N}_1$ where N_1 is a realization of the nerve of ω by a triangulation and follow it by the simplicial map $g: \tilde{N}_1 \to \tilde{N}$ generated by the natural bijection of the vertices of the complexes N_1 and N. That is, we set $f = g \circ f_1$. Such a map f will be called a *generalized barycentric map*. These maps are, in fact, defined by the same formula (6) as barycentric maps and if

$$\varphi_{i_0}(x) > 0, \ldots, \varphi_{i_r}(x) > 0, \qquad \sum \{\varphi_{i_k}(x): k = 0, \ldots, r\} = 1$$

then

$$f(x) \in T^r = |e_{i_0} \ldots e_{i_r}|$$

and the numbers $\{\varphi_{i_0}(x), \ldots, \varphi_{i_k}(x)\}$ are barycentric coordinates of the point $f(x)$ in the simplex T^r.

3.3. The Approximation Theorem.

A corollary of the canonical map theorem (Theorem 5) is the approximation theorem below.

Theorem 6. *Let $g: X \to I^m$ be a continuous map of an n-dimensional normal space X to an m-dimensional cube $I^m \subset \mathbb{R}^m$, $n \leqslant m$. Then, for any $\varepsilon > 0$, the following assertions hold*

1. *There exists a continuous map $f: X \to \tilde{N}$ into an n-dimensional polyhedron $\tilde{N} \subset I^m$ which is ε-close to g.*

2. *If $m \geqslant 2n + 1$ and if ω is any finite open cover of X, then we can suppose, in addition, that f is the canonical ω-map into $\tilde{N} = \tilde{N}(\omega)$ (and that there exists a subcomplex N' of the nerve N such that $f(x) = \tilde{N}'$).*

3. *Furthermore, let $\mathbb{R}_1^n$ be an arbitrary n-dimensional plane in $\mathbb{R}^m$ where $m \geqslant 2n + 1$. Then we can impose the additional requirement on the approximating map f that the set $\overline{f(X)}$ not intersect $\mathbb{R}_1^n$.*

Proof. Cover the cube I^m by a finite number of balls V_j of diameter less than $\varepsilon/2$. Choose an open cover $\omega_0 = \{O_1, \ldots, O_s\}$ of X finer than $\{g^{-1}V_j\}$ (and ω, if ω is given). Choose vertices $e_1, \ldots, e_s$ in the cube I^m in general position which are such that $\rho(e_i, g(O_i)) < \varepsilon/2$. If $m \geqslant 2n + 1$ and $\mathbb{R}_1^n$ a prescribed n-dimensional plane in $\mathbb{R}^m$, then we choose the points $e_1, \ldots, e_s$ subject to the additional condition that no n-dimensional plane generated by any $n + 1$ of the points e_1, ..., e_s meets $\mathbb{R}_1^n$. We then take N to be the geometrical realization of the nerve of the cover ω_0 with vertices $e_1, \ldots, e_s$ and f to be the canonical (generalized barycentric) map of Theorem 5 (Remark 2).

3.4. Theorems About ω-Maps, ε-Maps and ε-Shifts

Theorem 7. *A normal space X has dimension $\dim X \leqslant n$ if and only if for every finite open cover ω there exists an ω-map of X to a polyhedron of dimension less than or equal to n.*

The necessity immediately follows from Theorem 5. Sufficiency follows from the fact that, for every ω-map $f: X \to Y$, there exists an open cover u of Y whose preimage $f^{-1}u$ refines the cover ω.

It follows from Proposition 10 that Theorem 7 includes the following results.

Theorem 7_0. *A metric compactum X has dimension $\dim X \leqslant n$ if and only if, for each $\varepsilon > 0$, there exists an ε-mapping of X to a polyhedron of dimension less than or equal to n.*

Furthermore, the following result holds.

Theorem 8. *A compact subset X of euclidean space $\mathbb{R}^m$ has dimension $\dim X \leqslant n$ if and only if, for any $\varepsilon > 0$, we can map X by an ε-shift onto a polyhedron $\tilde{N} \subset \mathbb{R}^m$ of dimension less than or equal to n.*

Sufficiency follows from Theorem 7_0, necessity from Theorem 6 (take $f: X \to \tilde{N}$ to be a map approximating the identity map and, if $f(X) \neq \tilde{N}$, apply the process of sweeping out principal simplices not covered by the set $f(X)$).

Remark 3. The characterization of the dimension of metric compacta given in theorems 7_0 and 8 by ε-shifts and ε-maps cannot be extended to spaces with a countable base. In §7 we will cite an example constructed by Sitnikov of a two-dimensional set $X \subset \mathbb{R}^3$ which, for any $\varepsilon > 0$, admits ε-shifts into a one-dimensional polyhedron.

§4. The Nöbeling-Pontryagin Theorem

Theorem 9. *Every space X with a countable base and dimension $\dim X = n$ is homeomorphic to a subspace of the cube I^{2n+1}.*

The proof proceeds in three steps.

I. It follows from Theorem 6 that, for a given cover ω of the space X, the set $C^{(\omega)}$ of all ω-maps of X to the cube I^{2n+1} is everywhere dense in the space $C(X, I^{2n+1})$ of all continuous maps $f: X \to I^{2n+1}$.

II. The set $C^{(\omega)}$ is open in $C(X, I^{2n+1})$. Indeed, suppose that $f_0 \in C^{(\omega)}$. There exists an open cover v of I^{2n+1} such that $f^{-1}v$ is a refinement of ω. Taking $\varepsilon > 0$

126 V.V. Fedorchuk

small enough so that 5ε is the Lebesgue number of the cover v, it is not difficult to see that an ε-neighbourhood of f_0 lies in $C^{(\omega)}$. We remark that the assertion of this step is true for any X and the cube I^{2n+1} can be replaced by any (metric) compactum.

III. Now we choose a refining sequence of finite open covers

$$\omega_1, \omega_2, \ldots, \omega_k, \ldots$$

of X. The sets $C^{(\omega_k)}$ are open and everywhere dense in the complete space $C(X, I^{2n+1})$. Therefore, their intersection is nonempty and each $f \in \bigcap \{C^{(\omega_k)}: k = 1, 2, \ldots\}$ will be a homeomorphism.

The Nöbeling-Pontryagin theorem gives a topological characterization of subsets of Euclidean space.

In order that a topological space X be homeomorphic to a subset of some Euclidean space $\mathbb{R}^n$ it is necessary and sufficient that X be a finite-dimensional normal space with a countable base.

The Nöbeling-Pontryagin theorem can be strengthened.

Theorem 10. *If X is a space with a countable base and* $\dim X = n$, *the set of homeomorphisms of X into compacta lying in the set I_n^{2n+1} of points of the cube I^{2n+1} with no more than n rational coordinates is everywhere dense in the space $C(X, I^{2n+1})$.*

The proof is a refinement of that of Theorem 9. By Theorem 6, it is possible to arrange that the image $f(X)$ lies in a compactum which does not meet a countable number of preassigned n-dimensional planes $\mathbb{R}_1^n, \mathbb{R}_2^n, \ldots, \mathbb{R}_k^n, \ldots$. It remains to observe that I_n^{2n+1} is obtained from I^{2n+1} by subtracting a countable number of planes of the form

$$\mathbb{R}^n \begin{pmatrix} i_1, \ldots, i_{n+1} \\ r_1, \ldots, r_{n+1} \end{pmatrix},$$

each consisting of all points $x = (t_1, \ldots, t_{2n+1}) \in \mathbb{R}^{2n+1}$ which satisfy the $n + 1$ equations

$$t_{i_1} = r_1, \ldots, t_{i_{n+1}} = r_{n+1}$$

where $r_1, \ldots, r_{n+1}$ are rational numbers.

We can draw a very important corollary from Theorem 10. We let $C\begin{pmatrix} i_1 \cdots i_m \\ r_1 \cdots r_m \end{pmatrix}$ be the set of all points of the plane $\mathbb{R}\begin{pmatrix} i_1 \cdots i_m \\ r_1 \cdots r_m \end{pmatrix}$ all coordinates t_i of which, except $t_{i_1} = r_1, \ldots, t_{i_m} = r_m$, are irrational. This set is naturally homeomorphic to the $(m + 1)^{st}$ power of the set of irrational numbers on the line and, therefore, is inductively zero-dimensional. Furthermore, the set $C\begin{pmatrix} i_1 \cdots i_m \\ r_1 \cdots r_m \end{pmatrix}$ is closed in the set J_m^{2n+1} of all points of $\mathbb{R}^{2n+1}$ having exactly m rational coordinates and

$$J_m^{2n+1} = \bigcup C\begin{pmatrix} i_1 \cdots i_m \\ r_1 \cdots r_m \end{pmatrix}$$

where the union runs over all combinations $i_1, \ldots, i_m$ of m natural numbers not exceeding $2n + 1$ and all combinations $r_1, \ldots, r_m$ of rational numbers. According to the sum theorem for zero-dimensional sets (Proposition 9, Chap. 1) we have ind $J_m^{2n+1} = 0$. Set

$$Y_n^{2n+1} = J_0^{2n+1} \cup J_1^{2n+1} \cup \cdots \cup J_n^{2n+1}.$$

Then, by Proposition 10 of Chapter 1 we have

$$\text{ind } Y^{2n+1} \leqslant n. \tag{8}$$

We remark that Y_n^{2n+1} consists of all points of the space $\mathbb{R}^{2n+1}$ having no more than n rational coordinates. Therefore $I_n^{2n+1} = I^{2n+1} \cap Y_n^{2n+1}$ and, hence,

$$\text{ind } I_n^{2n+1} \leqslant n. \tag{9}$$

Theorem 11. *If X is a space with a countable base, then*

$$\text{ind } X \leqslant \dim X. \tag{10}$$

Indeed, suppose that dim $X \leqslant n$. Then, according to Theorem 10, the space X is homeomorphic to a subset of I_n^{2n+1}. Hence,

$$\text{ind } X \leqslant \text{ind } I_n^{2n+1} \leqslant n.$$

§ 5. The Characterization of Dimension by Means of Essential Maps and Partitions

5.1. The Essential Map Theorem. We begin with an elementary fact from homotopy theory.

The Mushroom Lemma. *Let A be a closed subset of a normal space X and let f_0 and f_1 be two homotopic maps of A to the sphere S^n. If f_1 can be extended to a map $g_1 \colon X \to S^n$, then f_0 can be extended to a map $g_0 \colon X \to S^n$ in such a way that g_0 and g_1 are homotopic.*

The lemma remains true if S^n is replaced by any absolute neighbourhood retract Y. The map $F \colon \Gamma \to Y$ defined on the "mushroom" $\Gamma = (A \times [0, 1]) \cup X \times \{1\}$ by the formula:

$$F(x, t) = \begin{cases} f_t(x) \colon x \in A \\ g_1(x) \colon t = 1 \end{cases}$$

extends to a map $G \colon O\Gamma \to Y$ of a neighbourhood of Γ in the product $X \times [0, 1]$. There exists a neighbourhood $OA \subset X$ such that $OA \times [0, 1] \subset O\Gamma$ (see Fig. 9). Let $\varphi \colon X \to [0, 1]$ be a function such that $\varphi(A) = 0$, $\varphi(X \setminus OA) = 1$. We then put $g_0(x) = G(x, \varphi(x))$.

The proposition below follows from the mushroom lemma.

Proposition 13. *Let $f = f_0$ be an essential map of a normal space X to a closed ball $\bar{T}^n$ with boundary S^{n-1} and $A = f^{-1}S^{n-1}$. Let $f_1 \colon X \to \bar{T}^n$ be a continuous map*

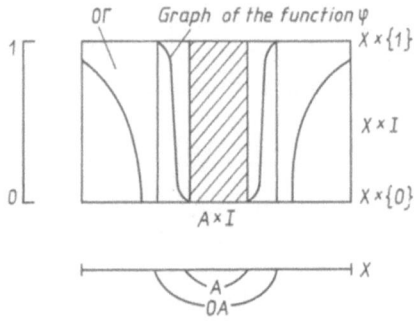

Fig. 9

for which $f_1^{-1} S^{n-1} \supset A$. *Then if the maps* $f|_A$ *and* $f_1|_A$ *are homotopic as maps into the sphere* S^{n-1}, *the map* $f_1 \colon X \to \bar{T}^n$ *is essential*.

From Proposition 13, in turn, we easily obtain the following result.

Proposition 14. *If* $f \colon X \to \bar{T}^n$ *is an essential map, then for each face* $\bar{T}^k$ *of the simplex* $\bar{T}^n$ *the map* $f \colon f^{-1}\bar{T}^k \to \bar{T}^k$ *is also essential*.

Theorem 12. *In order that a normal space* X *have dimension* $\dim X \geq n$ *it is necessary and sufficient that* X *map essentially onto an n-dimensional closed simplex* $\bar{T}^n$.

Proof. Necessity. From Theorem 7 about ω-maps and Theorem 5 about canonical maps, it follows that there exists a map f of the space X to a polyhedron $\tilde{K}$ of dimension greater than or equal to n such that each principal simplex is essentially covered. At the same time, the identity map of each principal simplex $T \in K$ can be extended to a simplicial map $g \colon \tilde{K} \to \bar{T}$ if we map all vertices which are not contained in T to some vertex of $\bar{T}$. Then the map $g \circ f$ is essential and applying Proposition 14 completes the proof.

Sufficiency. Let $f \colon X \to \bar{T}^n$ be an essential map to the ball $\bar{T}^n$. By using Proposition 13, it is not difficult to show that if $g \colon X \to \bar{T}^n$ is ε-close to f, then $g(X) \supset \bar{T}_0^n$ where $\bar{T}_0^n$ and $\bar{T}^n$ are concentric balls with radii r_0 and r, respectively, and $\varepsilon < \min\{r - r_0, r_0\}$. Then it follows from the approximation theorem (Theorem 6) that $\dim X \geq n$.

The existence theorem for essential maps has the following equivalent form.

Theorem 12. *A normal space* X *has dimension* $\dim X \leq n$ *if and only if each continuous map* f_A *of a closed set* $A \subset X$ *to the sphere* S^n *can be extended to a continuous map* $f_X \colon X \to S^n$.

5.2. Theorems About Partitions

Theorem 13. *A normal topological space* X *has dimension* $\dim X \leq n$ *if and only if for every collection of pairs*

$$(A_1, B_1), \ldots, (A_{n+1}, B_{n+1})$$

of disjoint closed subsets of X there exist sets C_i separating A_i and B_i and having empty intersection.

Proof. Necessity. Given a collection of pairs (A_i, B_i), construct a map $f: X \to Q^{n+1} = [-1, 1]^{n+1}$ such that $f(A_i) \subset {}^{-1}Q_i^n$ and $f(B_i) \subset {}^1Q_i^n$ where

$$^\varepsilon Q_i^n = \{(t_j) \in Q^{n+1}: t_i = \varepsilon\}.$$

This map is inessential by Theorem 12. Therefore, there exists a map $g: X \to S^n$ sweeping out to the boundary. Then $C_i = g^{-1}({}^0Q_i^n)$ will be the desired partitions.

Sufficiency. If $\dim X \geqslant n + 1$, then upon taking an essential map $f: X \to Q^{n+1}$ we obtain pairs

$$(A_i = f^{-1}({}^{-1}Q_i^n), B_i = f^{-1}({}^1Q_i^n))$$

of sets which cannot be separated by sets with empty intersection.

We deduce from the theorem above a geometric proposition which we will need in what follows.

Lemma 2. *Let C_1 be a set in Q^n which separates the opposite faces ${}^{-1}Q_1^{n-1}$ and ${}^1Q_1^{n-1}$ and let $u = \{U_1, \ldots, U_s\}$ be an open cover of C_1 such that no element U_i intersects opposite faces of the cube Q^n. Then $\operatorname{ord} u \geqslant n$.*

Proof. Suppose that $\operatorname{ord} u \leqslant n - 1$; we construct a barycentric map $f: C_1 \to N(u)$. From the condition on the element u it follows that the pairs

$$(f({}^{-1}Q_i^{n-1} \cap C_1), f({}^1Q_i^{n-1} \cap C_1)), \qquad i = 2, 3, \ldots, n,$$

are disjoint. Since $\dim N(u) \leqslant n - 2$, there exist sets D_i separating $f({}^{-1}Q_i^{n-1} \cap C_1)$ and $f({}^1Q_i^{n-1} \cap C_1)$ and having empty intersection. The sets $f^{-1}(D_i)$ separate ${}^{-1}Q_i^{n-1} \cap C_1$ and ${}^1Q_i^{n-1} \cap C_1$ in C_1 and extend to sets C_i in Q^n which separate ${}^{-1}Q_i^{n-1}$ and ${}^1Q_i^{n-1}, i = 2, 3, \ldots, n$. Then $C_1, C_2, \ldots, C_n$ will separate opposite faces of the cubes Q^n and have empty intersection. It easily follows that the identity map of Q^n to itself is inessential.

§6. The Sum Theorem and its Corollaries. The Equality ind X = Ind X = dim X for Spaces with a Countable Base

6.1. The Sum Theorem for the Dimension dim

Definition 9. If A is a subset of a normal space X we will write

$$\operatorname{rd}_X A \leqslant n$$

(and say that the *relative dimension of A in X* is less than or equal to n) if $\dim F \leqslant n$ for subset F of A which is closed in X.

Theorem 14. *If there exists a closed subset F of a normal space X such that $\dim F \leqslant n$ and $\operatorname{rd}_X(X \backslash F) \leqslant n$, then $\dim X \leqslant n$.*

Proof. By Theorem 12′ it is necessary to extend to X the continuous map $f_A: A \to S^n$ of the closed set A. By Theorem 12′ again, there exists an extension $g: F \to S^n$ of the map $f = f_A: A \cap F \to S^n$. The continuous map $h: A \cup F \to S^n$ which is equal to f_A on A and g on F can be extended to the closure $\bar{U}$ of a neighbourhood $U \supset A \cup F$. Since $\dim(X \setminus U) \leqslant n$, the continuous map $h: \mathrm{Bd}\, U \to S^n$ can be extended to a map $h_1: X \setminus U \to S^n$. Then the map f_X which is equal to h on $\bar{U}$ and h_1 on $X \setminus U$ is the desired extension of f_A.

Theorem 15 (The Sum Theorem). *If a normal space X is a union of closed subsets X_i, $i = 1, 2, \ldots$, of dimension $\dim X_i \leqslant n$, then $\dim X \leqslant n$.*

For a finite number of X_i, Theorem 15 follows immediately from Theorem 14. The general case is much harder: the map $f_A: A \to S^n$ successively extends to closed neighbourhoods of the partial unions $A \cup X_1 \cup \cdots \cup X_k$.

The equality

$$\dim \mathbb{R}^n = n \tag{11}$$

follows from Theorem 15 and the equality (5) (that is, $\dim P^n = n$).

6.2. Uryson's Identity ind X = Ind X = dim X

Lemma 3. *Let X be a Lindelof space of dimension ind $X \leqslant n$ and w an open cover of X. Then there exists a countable disjoint collection of open sets*

$$u = \{U_1, U_2, \ldots\}$$

and a countable collection of closed sets

$$\alpha = \{A_1, A_2, \ldots\}$$

satisfying the conditions
1°. *the collections u and α are refinements of w.*
2°. *ind $A_k \leqslant n - 1$ for any $i = 1, 2, \ldots$.*
3°. $X = \left(\bigcup_i U_i\right) \cup \left(\bigcup_i A_i\right).$

Proof. By hypothesis, there exists a refinement $v = \{V_1, V_2, \ldots\}$ of ω with ind Bd $V_i \leqslant n - 1$. We put $A_i = \mathrm{Bd}\, V_i$ and $U_i = V_i \setminus \bigcup \{V_k: k < i\}$.

Theorem 16. *If X is a strongly paracompact space, then*

$$\dim X \leqslant \mathrm{ind}\, X. \tag{12}$$

We quote the proof for a Lindelof space (which is, in fact, the same as in the general case). We proceed by induction on $n = \mathrm{ind}\, X$. Although it is possible to begin the induction with $n = -1$, we have already analyzed the case $n = 0$ in Proposition 7 of Chapter 1. Let ind $X = n$ and $\omega = \{O_1, \ldots, O_s\}$ be an open cover of X. Choose systems u and α as in Lemma 3 and set $A = \bigcup \{A_i: i = 1, 2, \ldots\}$. By the induction assumption, $\dim A_i \leqslant n - 1$ and hence, by the sum theorem, $\dim A \leqslant n - 1$.

The set $F = X \backslash \bigcup \{U_i : i = 1, 2, \ldots\}$ is closed and contained in A by formula 3°
so that $\dim F \leqslant n - 1$. The cover $\omega_F = \{F \cap O_1, \ldots, F \cap O_s\}$ of F has a combinatorial refinement by a closed cover $\{F_1, \ldots, F_s\}$ of multiplicity less than or equal
to n. By the thickening lemma (§ 2, Chapter 1) each set F_i can be replaced by a
neighbourhood $V_i \subset O_i$ so that the collection

$$v = \{V_1, \ldots, V_s\}$$

has multiplicity less that or equal to n. Then $u \cup v$ is a cover of X which has
multiplicity less than or equal to $n + 1$ and which is a refinement of ω.

The proof that

$$\dim X \leqslant \operatorname{Ind} X \tag{13}$$

for any normal space X proceeds along the same lines as the proof of Theorem
16, but is simpler.

Combining inequalities (10) and (12) shows that

$$\operatorname{ind} X = \dim X \tag{14}$$

for any space X with a countable base.

To complete the proof of Uryson's theorem, we need to verify that

$$\operatorname{Ind} X \leqslant \operatorname{ind} X \tag{15}$$

for any space X with a countable base.

We proceed by induction on $n = \operatorname{ind} X$. Suppose that F is closed in X and O_F
is any neighborhood. Let O_{1F} be a neighborhood whose closure lies in O_F. Apply
Lemma 3 to the cover $\omega = \{O_F, X \backslash \overline{O_{1F}}\}$. Divide u into two subcollections

$$u_1 = \{U \in u : U \cap O_{1F} \neq \varnothing\}$$

and

$$u_2 = \{U \in u : U \cap O_{1F} = \varnothing\}.$$

We put $V_1 = \tilde{u}_1$, $V_2 = \tilde{u}_2$ and $O_{2F} = O_{1F} \cup V_1$. Then $O_{2F} \cap V_2 = \varnothing$ and, according to formula 3°,

$$\operatorname{Bd} O_{2F} \subset A = \bigcup \{A_i : i = 1, 2, \ldots\}.$$

From the sum theorem and equality (14), it follows that

$$\operatorname{ind} A = \dim A \leqslant n - 1.$$

Therefore, by the induction assumption, $\operatorname{Ind} \operatorname{Bd} O_{2F} \leqslant n - 1$. This, together with
the inclusion $O_{2F} \subset O_F$ gives the inequality $\operatorname{Ind} X \leqslant n$.

This completes the proof of Uryson's identity

$$\operatorname{ind} X = \operatorname{Ind} X = \dim X \tag{16}$$

for spaces with a countable base.

6.3. The Theorems on Decompositions, Products, Universal Compacta, and Compactifications

Theorem 17. *In order that a nonempty space with a countable base have dimension* dim $X \leqslant n$, *it is necessary and sufficient that X can be represented as a union of $n + 1$ zero-dimensional sets.*

Proof. Sufficiency follows from inequality (14) of Chapter 1. To establish necessity, it is sufficient to take a countable base $\mathscr{B} = \{V_i\}$ consisting of sets V_i in X with dim Bd $V_i = $ ind Bd $V_i \leqslant n - 1$, to set $X^1 = \bigcup \{\text{Bd } V_i : i = 1, 2, \ldots\}$ and $X_0 = X \setminus X^1$ and to apply the induction hypothesis to X^1.

Theorem 18. *If X and Y are spaces with a countable base, then*

$$\dim(X \times Y) \leqslant \dim X + \dim Y. \qquad (17)$$

This theorem follows from the identity ind $X = \dim X$, the sum theorem, and the following, easily verified identity:

$$\text{Bd}(U \times V) = (\text{Bd } U \times \bar{V}) \cup (\bar{U} \times \text{Bd } V).$$

Remark 4. Inequality (17) cannot be strengthened to an equality. In §4 of Chapter 5, we will cite examples of metric compacta X and Y such that

$$\dim(X \times Y) < \dim X + \dim Y.$$

Such compacta necessarily have dimension at least 2 since it is possible to show that if dim $Y > 0$, then

$$\dim(X \times Y) \geqslant \dim X + 1. \qquad (18)$$

Inequality (18) does not hold without the assumption on compactness. Indeed, it is not difficult to show that the space l_2^r of §2 of Chapter 1 is one-dimensional. At the same time it is homeomorphic to its square. Therefore,

$$\dim(l_2^r \times l_2^r) = \dim l_2^r = 1.$$

In §4, we saw (Theorem 10) that any space with a countable base and dimension dim $X \leqslant n$ can be topologically imbedded into a compactum lying in I_n^{2n+1}. It was shown that ind $I_n^{2n+1} \leqslant n$. Hence, dim $I_n^{2n+1} \leqslant n$ and I_n^{2n+1} itself is embedded in a compactum C_n lying in I_n^{2n+1}. Thus, the following holds.

Theorem 19. I. *For each integer $n \geqslant 0$ there exists an n-dimensional compactum C_n containing a topological image Z of each space X with a countable base and dimension* dim $X \leqslant n$.
  II. *The closure $\bar{Z}$ is an n-dimensional compactum which is a compactification of X.*

In the next section we will again return to universal compacta in Euclidean spaces.

§ 7. On the Dimension of Subsets of Euclidean Space

7.1. Sets of Dimension n in Euclidean Space $\mathbb{R}^n$

Theorem 20. *A set $X \subset \mathbb{R}^n$ has dimension n if and only if X has interior points.*

Sufficiency follows from the fact that every nonempty open subset of $\mathbb{R}^n$ contains an n-dimensional simplex $\bar{T}^n$.

Necessity. Suppose the complement of X is everywhere dense in $\mathbb{R}^n$ and $u = \{U_1, \ldots, U_s\}$ is an open cover of the space X. There exist open subsets V_i of $\mathbb{R}^n$ cutting out the sets U_i on X. Set $v = \{V_1, \ldots, V_s\}$ and $V = \tilde{v}$. Now construct a sequence of n-dimensional cubes Q_i^n in $\mathbb{R}^n$ with edges parallel to the coordinate axes satisfying the following conditions

1°. The interiors of different cubes Q_i^n and Q_j^n do not intersect.
2°. $\bigcup \{Q_i^n : i = 1, 2, \ldots\} = V$.
3°. The collection $q = \{Q_1^n, Q_2^n, \ldots\}$ is locally finite in V and refines v.

Let $S_i^{n-1} = \text{Bd } Q_i^n$ and $S = \bigcup \{S_i^{n-1} : i = 1, 2, \ldots\}$. Then $\dim S = n - 1$. Finally, construct a u-map $f : X \to S$ which sweeps the set $X \cap \text{Int } Q_i^n$ out to the boundary S_i^{n-1} by means of central projection from a point $p_i \in \text{Int } Q_i^n$ not belonging to X.

7.2. Universal Menger Compacta M_n^m in the Euclidean Space $\mathbb{R}^n$.

We now detail a construction, special cases of which include the Cantor perfect set, the Sierpinski carpet, and the universal Menger curve. The n-dimensional skeleton $\text{Sk}_n Q^m$ of an m-dimensional cube Q^m is the union of all n-dimensional faces. Divide the standard Euclidean cube into 3^m identical cubes ${}^1 Q_i^m$ called cubes of the first rank. Let K_1 denote the union of all cubes of the first rank which meet the n-dimensional skeleton $\text{Sk}_n I^m$. Do the same to each of the first rank cubes in K_1 to obtain a compact set K_2. The intersection $\bigcap \{K_i : i = 1, 2, \ldots\}$ is compact and is denoted M_n^m. It is of dimension n because, for each $\varepsilon > 0$, it admits an ε-map into an n-dimensional polyhedron $\text{Sk}_n K_i$ for sufficiently large i.

It is clear that M_0^1 is the Cantor perfect set, M_1^2 the Sierpiński carpet and M_1^3 the universal Menger curve.

Theorem 21. *The compactum M_{n-1}^n topologically contains every closed set $X \subset \mathbb{R}^n$ of dimension $\dim X \leqslant n - 1$.*

The proof for compact sets was, in fact, "depicted" in Fig. 4. In the general case, it is necessary to count sufficiently precisely. The idea of the proof is similar to the proof of Theorem 20, except that the "sweeping out" of $X \cap \text{Int } Q_i^n$ to the boundary S_i^{n-1} which is fixed outside of $\text{Int } Q_i^n$ is replaced by a "sweeping out" (topological deformation) which is fixed outside of a small neighbourhood of Q_i^n.

Theorem 22. *The compactum M_n^{2n+1} topologically contains every set $X \subset \mathbb{R}^{2n+1}$ of dimension $\dim X \leqslant n$; that is, it is universal for n-dimensional separable metric spaces.*

To prove this it is sufficient to imbed the space I_n^{2n+1} of § 4 into M_n^{2n+1}. But this was essentially done in the proof of Theorem 10, where X was imbedded in a compactum lying in I_n^{2n+1} and, consequently, in a space obtained from I^{2n+1} by throwing out some neighborhoods $O_{r_1}^{i_1} \ldots {}_{r_{n+1}}^{i_{n+1}}$ of the planes $\mathbf{R}^n({}_{r_1}^{i_1}, \ldots, {}_{r_{n+1}}^{i_{n+1}})$. For suitable choices of neighborhoods $O_{r_1}^{i_1} \ldots {}_{r_{n+1}}^{i_{n+1}}$, this also gives us a space homeomorphic to M_n^{2n+1} (recall that the main step in constructing the compactum M_n^m was the transition from the polyhedra K_i to K_{i+1} by throwing out of K_i the "rectangular" neighbourhoods of $(m - n - 1)$-dimensional polyhedra dual to $\mathrm{Sk}_n K_i$). The following question is still open.

Question. *Can every space $X \subset \mathbb{R}^m$ of dimension* dim $X \leqslant n$ *be imbedded in M_n^m?*
This question is equivalent to the following problem due to Aleksandrov.

Does every space $X \subset \mathbb{R}^m$ of dimension dim $\leqslant n$ have an n-dimensional compactification which imbeds in $\mathbb{R}^m$?

The equivalence of these two questions follows from the following theorem proved by Shtanko.

Theorem 23. *The compactum M_n^m topologically contains every compactum $X \subset \mathbb{R}^m$ of dimension* dim $X \leqslant n$.

7.3. Separation of $\mathbb{R}^n$ by Subsets. We will say that a closed subset $F \subset X$ *separates* the space X if it separates two non-empty sets; that is, if $X \backslash F$ is a disjoint union $U_1 \cup U_2$ of nonempty open sets.

Theorem 24. *A domain of $\mathbb{R}^n$ and, in particular $\mathbb{R}^n$ itself, cannot be separated by a subset of dimension less than or equal to $n - 2$.*

First note that no compactum of dimension less than or equal to $n - 2$ separates $\mathbb{R}^n$: otherwise, the proposition about the existence of such compacta would immediately imply that ind $\mathbb{R}^n \leqslant n - 1$. We claim that this implies that every compactum F which separates a closed ball B^n has dimension at least $n - 1$. Indeed, take any point $0 \in \mathrm{Int}\, B^n \backslash F$ and consider inversion $h: \mathbb{R}^n \backslash \{0\} \to \mathbb{R}^n \backslash \{0\}$ with respect to $S^{n-1} = \mathrm{Bd}\, B^n$. We obtain a compactum $\Phi = F \cup h(F)$ which separates $\mathbb{R}^n$ and which is the union of two homeomorphic compacta F and $h(F)$. Finally, upon covering the connected open set $U \subset \mathbb{R}^n$ with a countable number of open balls O_i, we find that for a closed set $F \subset U$ of dimension less than or equal to $n - 2$, the set $O_i \backslash F$ is connected. Therefore, the set $U \backslash F$ is connected as the support of the special system $\{O_i \backslash F : i = 1, 2, \ldots\}$ consisting of connected sets (see article I, §8)

Definition 10. A space X of dim $X = n$ is called an *n-dimensional Cantor manifold* if it cannot be separated by a subset of dimension less than or equal to $n - 2$.

By taking Theorem 20 into account, we can reformulate Theorem 24 in the following equivalent manner.

Theorem 24′. *Every domain $U \subset \mathbb{R}^n$ is an n-dimensional Cantor manifold.*

Theorem 24, obtained by Menger and Uryson, was strengthened in 1929 by Mazurkiewicz.

Theorem 25. *A subset X of a domain $U \subset \mathbb{R}^n$ which has dimension less than or equal to $n - 2$ does not cut U; that is, for any pair of points x, $y \in U \backslash X$, there exists a continuum $K \subset U \backslash X$ joining x and y.*

Proof. It suffices to prove that a set X which cuts the cube Q^n between opposite faces $^{-1}Q_1^{n-1}$ and $^1Q_1^{n-1}$ has dimension greater than or equal to $n - 1$. For this it suffices to prove that ord $v \geqslant n$ for each open cover $v = \{V_1, \ldots, V_s\}$ of X, no element of which meets the opposite faces of the cube Q^n. By Čech's lemma in §4 of Chapter 1, the cover v can be extended to a similar collection $u = \{U_1, \ldots, U_s\}$ of open subsets of Q^n with the same restrictions to the faces of Q^n. The set $U_1 \cup \cdots \cup U_s$ contains a set C_1 which separates $^{-1}Q_1^{n-1}$ and $^1Q_1^{n-1}$. By Lemma 2, we have ord $v =$ ord $u \geqslant n$.

A set, any two points of which can be joined by a continuum, is called a *semicontinuum*. In some cases it is necessary to specify the minimal dimension of the continuum connecting the given points. It is easy to show that any two points x and y of a semicontinuum $X \subset \mathbb{R}^n$ can be connected by a continuum K of dimension less than $n - 1$. In the general case, it is not possible to lower this dimension, A.V. Ivanov established the following.

Examples. There exists a one dimensional set $A \subset \mathbb{R}^3$ whose complement $X = \mathbb{R}^3 \backslash A$ is a semicontinuum by Menger's theorem, but which contains points x and y which cannot be joined by a one-dimensional continuum.

For later use, we will need the following.

Sierpiński's Theorem. *No continuum can be represented as a disjoint union of countably many nonempty closed sets.*

Suppose that such a representation is possible: $X = \bigcup \{X_i : i = 1, 2, \ldots\}$. Take disjoint neighbourhoods U_1 and U_2 of X_1 and X_2, respectively. Let K_1 be a component of the closure $\bar{U}_2$ which meets X_2. For an arbitrary continuum Y and an arbitrary closed subset Z, $\varnothing \neq Z \neq Y$, any component of Z meets Bd Z. Hence, the continuum K_1 meets some set X_i for $i > 2$ and, consequently, an infinite number of them. Continuing this process, we construct a decreasing sequence $K_1 \supset K_2 \supset \cdots$ of continua with the property that $K_i \cap X_i = \varnothing$ and, hence, $\bigcap \{K_i : i = 1, 2, \ldots\} = \varnothing$ which contradicts the compactness of X.

Sitnikov's Example. *There exists a two-dimensional subset X of $\mathbb{R}^3$ which, for any $\varepsilon > 0$, can be mapped by an ε-shift into a one-dimensional continuum.*

For a given natural number i let $\mathcal{K}$ be the collection of all cubes into which $\mathbb{R}^3$ is divided by the planes defined by the equations $t_j = n/i$ where $j = 1, 2, 3$, and n is an integer. The union of all edges of cubes in $\mathcal{K}_i$ will be denoted by A_i.

We put $B_1 = A_1$. For $i = 2, 3, \ldots$, we let B_i be a set obtained by a parallel translation of A_i which is such that B_i does not intersect $B_1 \cup \cdots \cup B_{i-1}$. Set $B = \bigcup \{B_i : i = 1, 2, \ldots\}$ and $X = I^3 \backslash B$. Since B is everywhere dense in $\mathbb{R}^3$ we have dim $X \leqslant 2$. On the other hand, $I^3 \backslash X$ is a countable disjoint union of the closed sets $I^3 \cap B_i$ and, consequently, cannot be a semicontinuum by Sierpinski's

136 V.V. Fedorchuk

theorem. It follows from Mazurkiewicz's theorem that dim $X \geqslant 2$. Consequently, dim $X = 2$.

Finally, for $i > \sqrt{3}/\varepsilon$, the set X admits an ε-shift to a polyhedron $B_i \cap I^3$. Indeed, choose a point q in a cube $Q^3 \in \mathcal{K}_i$ which is a lattice point of B_{i+1} and sweep the set $Q^3 \cap X$ out to the boundary $S^2 = \text{Bd } Q^3$ by means of central projection f_q from q. For a suitable choice of lattice points for different cubes $Q^3 \in K_i$, we obtain a map f which sweeps out to the two-dimensional faces of the cubes $Q^3 \in \mathcal{K}_i$ and which fixes points not covered by the set $f(X)$. Now any "two dimensional face" of $f(X) \cap Q^2$ can be swept out onto its one dimensional skeleton.

Composing these two sweeping outs gives the desired ε-shift.

Definition 11. A continuous map $f: X \to S^n$ is called *essential* if it is not homotopic to a constant map.

The following very deep and significant theorem was obtained in 1932 independently by Aleksandrov and Borsuk.

Theorem 26. *In order that a compact subset $F \subset \mathbb{R}^n$ separate $\mathbb{R}^n$ for $n \geqslant 2$, it is necessary and sufficient that F can be mapped essentially to the sphere S^{n-1}.*

Necessity follows immediately from the assertion below. The latter follows, in turn, from the essentialness of the identity map of a simplex (Theorem 9) and the mushroom lemma. The proof will be left to the reader.

Lemma 4. *A compact subset $F \subset \mathbb{R}^n$ separates the points x and y in $\mathbb{R}^n$ if and only if the maps $\pi_x: F \to S^{n-1}$ and $\pi_y: F \to S^{n-1}$ are not homotopic to one another.*

Here, $\pi_z: \mathbb{R}^n \setminus \{z\} \to S^{n-1}$ denotes the composition of central projection $\mathbb{R}^n \setminus \{z\} \to S_z^{n-1}$ to the sphere centered at the point z and parallel translation of S_z^{n-1} to S^{n-1}.

The proof of sufficiency is considerably more complicated; the reader can find a proof in [AP].

The depth of the Aleksandrov-Borsuk theorem is illustrated by the following corollary.

The Generalized Jordan Theorem. *If $h: S^{n-1} \to \mathbb{R}^n$ is an imbedding, the set $h(S^{n-1})$ separates $\mathbb{R}^n$.*

Indeed, it suffices to prove that the identity map $S^{n-1} \to S^{n-1}$ is essential. For, if we suppose that there exists a homotopy f_t between the identity map f_1 and the constant map f_0, then we obtain a contradiction to the essentialness of the identity map of the ball by letting $g: B^n \to S^{n-1}$ be the map sweeping the ball out to its boundary given by the equation

$$g(r, s) = f_r(s)$$

where r is the "radial" and s the "spherical" coordinate of a point $z \in B^n$.

Yet another theorem of Aleksandrov follows from Theorem 26.

Theorem 27. *If a compact subset $F \subset \mathbb{R}^n$ is the common boundary of at least two domains in $\mathbb{R}^n$, then it is an $(n-1)$-dimensional Cantor manifold.*

The proof uses the following elementary fact from homotopy theory which is based on the characterization of dimension in terms of extensions of maps to the sphere.

Lemma 5. *Let f and g be continuous maps of a compactum X to a sphere S^n. Let $X = A_1 \cup A_2$ where A_1 and A_2 are closed and $\dim(A_1 \cap A_2) \leqslant n - 2$. If the maps f and g are homotopic on A_1 and A_2, then they are also homotopic on X.*

Chapter 3
Dimension of General Spaces

§1. The Factorization Theorem and Spectral Decomposition of Compacta. Universal Compacta of Prescribed Weight and Dimension

1.1. Mardeshich's Factorization Theorem and Spectral Decomposition of Compacta

Theorem 1. *If $f: X \to Y$ is a continuous map of a compactum X to a compactum Y, there exists a compactum Z and continuous maps $g: X \to Z$ and $h: Z \to Y$ such that $f = h \circ g$, $\dim Z \leqslant \dim X$, and $wZ \leqslant wY$.*

Proof. There exists a collection $\mathscr{V}$ of open covers of Y of cardinality $|\mathscr{V}| = \tau = wY$ such that each open cover of Y has a refinement $v \in \mathscr{V}$. Each cover $f^{-1}v$ of X has an open refinement u_v of multplicity less than or equal to $n + 1$. The collection $\mathscr{U}_0 = \{u_v: V \in \mathscr{V}\}$ can be augmented to a collection $\mathscr{U}$, with the same cardinality τ, of open covers of X of multiplicity at most $n + 1$ in such a way that $\mathscr{U}$ is a basis of a uniformity. By the lemma about associated separated uniformities (see article I of this volume, §12), there exists a space $Z = Z_{\mathscr{U}}$ and a continuous map $g = g_{\mathscr{U}}: X \to Z$ onto Z such that the collection $g \# \mathscr{U} = \{g \# u: u \in \mathscr{U}\}$ is a basis of a separable uniformity on Z. Under our hypotheses, Z will be a compactum of weight $wZ \leqslant |\mathscr{U}| = \tau = wY$ and dimension less than or equal to n (the latter follows because each cover $g \# u = \{g \# U: U \in u\}$ has multplicity less than or equal to $n + 1$). It remains to construct the map h. The elements of the decomposition $\mathscr{R}_f = \{f^{-1}y: y \in Y\}$ have the form

$$\bigcap \{\operatorname{st}_{f_v^{-1}}x: v \in \mathscr{V}\}$$

and the elements of the decomposition $\mathscr{R}_g = \{g^{-1}z: z \in Z\}$ have the form

$$\bigcap \{\operatorname{st}_u x: u \in \mathscr{U}\}.$$

Since the collection $\mathscr{U}$ is a refinement of $f^{-1}\mathscr{V}$, the decomposition $\mathscr{R}_g$ is a refinement of $\mathscr{R}_f$. Therefore there exists a unique map $h: Z \to Y$ such that

$f = h \circ g$. From the equality $f = h \circ g$ and the quotient construction of g, it immediately follows that h is continuous. Theorem 1 is proved.

Another theorem due to Mardeshich follows from Theorem 1.

Theorem 2. *Every compactum X of dimension* $\dim X \leqslant n$ *is the limit of an inverse spectrum of metrizable compacta of dimension less than or equal to n.*

Proof. Imbed X in the Tikhonov cube I^τ of weight $\tau = wX$. Let $\mathscr{P}_\omega(\tau)$ be the set of all countable subsets of τ. The compactum X is a limit of the σ-spectrum $S = \{X_A, \pi_A^A : A \in \mathscr{P}_\omega(\tau)\}$ of metrizable compacta $X_A = p_A(X)$ where $\pi_{A'}^A = p_{A'}^A|$ X_A. We shall show that the set $\mathscr{A}$ of A for which $\dim X_A \leqslant n$ is cofinal in $\mathscr{P}_\omega(\tau)$. Suppose $A_0 \in P_\omega(\tau)$. By Theorem 1 there exists a metrizable compactum X_1 of dimension less than or equal to n and a map $g_1 : X \to X_1$ and $h_1 : X_1 \to X_{A_0}$ such that $\pi_{A_0} = h_1 \circ g_1$. According to the factorization lemma (see article I of this volume, §11) there exists a countable set $A_1 \supset A_0$ and a continuous map $f_1 : X_A \to X_1$ such that $g_1 = f_1 \circ \pi_A$. By repeatedly applying Theorem 1 and the factorization lemma, we construct a metrizable compactum X_k of dimension less than or equal to n, sets $A_k \in \mathscr{P}_\omega(\tau)$, and continuous maps $g_k : X \to X_k$, $h_k : X_k \to X_{A_{k-1}}$, $f_k : X_{A_k} \to X_k$ such that

1) $A_k \subset A_{k+1}$,
2) $\pi_A = h_{k+1} \circ g_{k+1}$, and
3) $g_k = f_k \circ \pi_A$.

Suppose that $A = \cup A_k$. Then the compactum X_A is a limit of the spectrum of compacta X_k and mappings $f_{k-1} \circ h_k : X_k \to X_{k-1}$. But $\dim X_k \leqslant n$. This implies that $\dim X_A \leqslant n$ because it immediately follows from the definition of the topology of the limit of an inverse spectrum that the assertion below is true.

Proposition 1. *If a compactum X is a limit of an inverse spectrum of compacta of dimension less than or equal to n, then* $\dim X \leqslant n$.

An analysis of the proof of Theorem 2 shows that we have actually proved a stonger assertion.

Theorem 2_0. *If a compactum X of dimension less than or equal to n can be expanded in a sigma-spectrum S of metrizable compacta, then there exists a cofinal, closed subspectrum S_0 of S consisting of compacta of dimension less than or equal to n.*

Theorem 2 can be extended to a theorem about spectral decomposition of metrizable compacta. Using a special version of Theorem 5 about canonical maps one can prove the following.

Theorem 3 (Fraudenthal). *Every n-dimensional metrizable compactum X is a limit of an inverse spectrum $S = \{\tilde{K}_i, \pi_i^{i+1} : i = 1, 2, \ldots\}$ of n-dimensional polyhedra $\tilde{K}_i$ with fixed triangulations K_i. In addition, each π_i^{i+1} is a simplicial map of K_{i+1} to a multiple barycentric subdivision K_i^* of K_i and every projection $\pi_i : X \to \tilde{K}_i$ essentially covers each principal simplex $T \in K_i$.*

It would seem that combining Theorems 2 and 3 would make it possible to show that any n-dimensional compactum can be expanded in a spectrum of n-dimensional polyhedra. But this is not so. Indeed, because every nowhere dense subset of an n-dimensional polyhedron has dimension less than or equal to $n - 1$, we obtain the following result.

Proposition 2. *If a compactum X is a limit of a spectrum of one-dimensional polyhedra, then*

$$\dim X = \operatorname{ind} X = \operatorname{Ind} X.$$

Therefore the one-dimensional compactum L of §2 is not a limit of an inverse spectrum of one-dimensional polyhedra. We also remark that the coincidence of the dimensions dim, ind, and Ind for a compactum does not guarantee that it is a limit of an inverse spectrum of polyhedra of the same dimension. In addition, Pasynkov has shown (see [1]) that there exist compacta which are not limits of inverse spectra of polyhedra or even locally connected compacta with projections which are "onto".

1.2. Universal Compacta and Compactifications of Prescribed Weight and Dimension

Proposition 3. *For any normal space X*

$$\dim X = \dim \beta X. \tag{1}$$

The assertion follows because each finite open cover $\{O_1, \ldots O_s\}$ of X extends to an open cover $\{O_1^\beta, \ldots, O_s^\beta\}$ of the compactification βX and enlarging open collections from everywhere dense subsets does not increase multiplicity.

Theorem 4 (Zarelya, Pasynkov). *For any infinite cardinal number τ and any integer $n \geqslant 0$, there exists a compactum Π_τ^n of weight τ and dimension n containing a homeomorphic image of every completely regular space X with weight at most τ and $\dim \beta X \leqslant n$. In particular, Π_τ^n contains every normal space of weight less than or equal to τ and dimension less than or equal to n.*

Proof. Tikhonov's theorem implies that there exist no more than 2^{2^τ} completely regular spaces X_α, $\alpha \in A$, of weight less than or equal to τ with $\dim \beta X_\alpha \leqslant n$. Suppose $Y = \oplus \beta X_\alpha$ is a discrete sum of Stone-Čech compactifications of the spaces X_α. By Proposition 3, $\dim \beta Y \leqslant n$.

For each α, there exists an imbedding $f_\alpha \colon X_\alpha \to I^\tau$ which extends to a continuous map $\bar{f}_\alpha \colon \beta X_\alpha \to \bar{I}^\tau$. There exists a map $f \colon \beta Y \to I^\tau$ coinciding with $\bar{f}_\alpha$ on each βX_α. According to Theorem 2, there exists a bicompact space Π_τ^n and maps $g \colon \beta Y \to \Pi_\tau^n$ and $h \colon \Pi_\tau^n \to I^\tau$ such that $f = h \circ g$, $\dim \Pi_\tau^n \leqslant n$ and $w\Pi_\tau^n \leqslant \tau$. The map $f|X_\alpha = f_\alpha$ is a homeomorphism onto its image and, hence, so is $g|X_\alpha$. The inequalities $\dim \Pi_\tau^n \geqslant n$ and $w\Pi_\tau^n \geqslant \tau$ are obvious. The last assertion of Theorem 4 follows from Proposition 3.

An immediate corollary of Theorem 4 is the following.

Theorem 5 (Sklyarenko). *If X is a completely regular space with weight τ and $\dim \beta X \leqslant n$, in particular, if X is a normal space of weight τ and dimension at most n, then X has a compactification cX of weight τ and dimension at most n.*

Proposition 4 (Vedenisov). *For every normal space X*

$$\operatorname{Ind} X = \operatorname{Ind} \beta X. \tag{2}$$

The proof is by induction on the dimension, and uses the isomorphism $\bar{A}^{\beta X} \cong \beta A$ for a closed subset A of a normal space and the following result which is of independent interest.

Lemma 1. *If C is a set which separates closed subsets A and B of a normal space X, then $\bar{C}^{\beta X}$ separates $\bar{A}^{\beta X}$ and $\bar{B}^{\beta X}$ in βX.*

A factorization theorem also holds for the dimension Ind.

Theorem 6 (Pasynkov). *Let $f: X \to Y$ be a continuous map of compacta and suppose F is a closed subset of X of dimension $\operatorname{Ind} F = n$. Then there exists a compactum Z and continuous maps $g: X \to Z$ and $h: Z \to Y$ such that $f = h \circ g$, $wZ \leqslant wY$ and $\operatorname{Ind} f(F) \leqslant n$.*

From this theorem and Proposition 4, we obtain the following theorem by arguing as in the proof of Theorem 4.

Theorem 7. *For every infinite cardinal number τ and integer $n \geqslant 0$ there exists a compactum Ψ_τ^n of weight τ and dimension $\operatorname{Ind} \Psi_\tau^n = n$ which contains a homeomorphic image of any completely regular space X with $wX \leqslant \tau$ and $\operatorname{Ind} \beta X \leqslant n$. In particular, Ψ_τ^n contains every normal space X of weight less than or equal to τ and dimension $\operatorname{Ind} X \leqslant n$.*

A corollary to Theorem 7 is the following.

Theorem 8. *Every completely regular space X with weight τ and $\operatorname{Ind} \beta X \leqslant n$, and, in particular, every normal space X of weight τ and dimension $\operatorname{Ind} X \leqslant n$, has a compactification cX of weight τ and dimension $\operatorname{Ind} cX \leqslant n$.*

Since $\dim \leqslant \operatorname{ind} \leqslant \operatorname{Ind}$ for compacta and $\dim I^n = \operatorname{Ind} I^n = n$, Theorem 7 implies the following result.

Corollary 1. *The compactum Ψ_τ^n is universal in the class of compacta X with weight less than or equal to τ and dimension $\dim X = \operatorname{ind} X = \operatorname{Ind} X = n$.*

§2. The Relation Between the Dimensions dim, ind, and Ind of Compacta. The Dimension of Subspaces

2.1. Aleksandrov's Problem About the Relations Between the Fundamental Dimensional Invariants dim, ind and Ind in the Class of Compacta. Aleksandrov's problem was formulated in 1935.

It follows from Theorem 15 of Chapter 2 that every strongly paracompact and, in particular, compact space X satisfies

$$\dim X \leqslant \operatorname{ind} X \leqslant \operatorname{Ind} X. \tag{3}$$

For an arbitrary compact space, it is impossible to say more. The first examples of compacta for which the dimensions do not coincide were constructed in 1949 by Lunts and Lokutsievskiĭ. We present here a significantly simpler example due to Lokutsievskiĭ.

Let $A = A(\omega_1)$ be the transfinite line of length ω_1; that is, the topologically unique, connected, linearly ordered, compactum which is locally metrizable everywhere except at one of its endpoints. At this point the compactum A necessarily has character ω_1. As a linearly ordered space, A is isomorphic to the lexicographic product of the well ordered set $W(\omega_1)$ of transfinites $<\omega_1$ on the semi-interval $[0, 1)$ augmented by a largest element ω_1.

Let D_1 and D_2 be disjoint countable dense subsets of the segment $I = [0, 1]$. Let $f_i: C \rightarrow I$, $i = 1, 2$, be a topological copy of the standard projection of the Cantor set C to the interval (see § 3, chap 1) for which the points not of multiplicity one are the points of D_i and only these. Let B_i denote the space obtained by attaching $A \times C$ to I by the map $f_i: \{\omega_1\} \times C \rightarrow I$. Let L denote the space obtained from the discrete union of the compacta B_1 and B_2 by identifying the copies of the segment $\{\omega_1\} \times I$ in each (see Fig. 10). This is Lokutsievskiĭ's compactum.

Every set separating the upper and lower "faces" of L contains either a piece of the vertical segment $\{\omega_1\} \times I$ or a "horizontal" piece homeomorphic to A (a standard such separating set is depicted in Fig. 10 by two broken lines and a single solid line). Therefore, $\operatorname{Ind} L \geqslant 2$.

On the other hand, it is obvious that $\operatorname{ind} B_1 = \operatorname{ind} B_2 = 1$. Therefore, $\dim B_1 = \dim B_2 = 1$ and, by the sum theorem, $\dim L = 1$. In addition, for compacta,

$$\operatorname{ind} X = 1 \Rightarrow \operatorname{Ind} X = 1. \tag{4}$$

Therefore, $\operatorname{Ind} B_1 = \operatorname{Ind} B_2 = 1$.

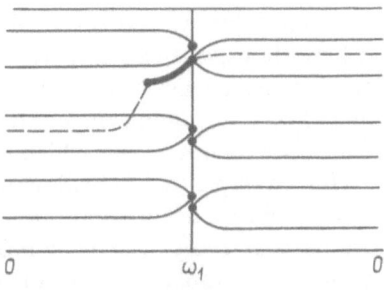

Fig. 10

Proposition 5. *If a normal space X is a union of nonempty closed subsets X_1 and X_2, then*

$$\operatorname{ind} X \leqslant \operatorname{ind} X_1 + \operatorname{ind} X_2 + 1 \tag{5}$$

$$\operatorname{Ind} X \leqslant \operatorname{Ind} X_1 + \operatorname{Ind} X_2. \tag{6}$$

The proof is by induction on the right hand sides of both inequalities using the following simple assertion.

Lemma 2. *Let A and B be disjoint closed sets in X and $A \cap X_1 \neq \varnothing \neq B \cap X_1$. Let F separate $A \cap X_1$ and $B \cap X_1$ in X_1. Then $F \cup X_2$ contains a set which separates A and B in X.*

To verify (6) we examine separately the case when the summands are zero-dimensional.

From the aforementioned inequality, $\operatorname{Ind} L \geqslant 2$, together with proposition 5 and property (4), it follows that

$$\operatorname{ind} L = \operatorname{Ind} L = 2.$$

Thus,

$$1 = \dim L < \operatorname{ind} L = \operatorname{Ind} L = 2.$$

Lokutsievskiĭ's example also shows that, in contrast to the dimension dim, the sum theorem does not hold for the inductive dimension, even in the case of two compact summands.

Another notable example of a compactum with unequal dimensions is Fedorchuk's example of a separable compactum Θ satusfying the first axiom of countability and having dimensions

$$\dim \Theta = 2, \qquad 3 \leqslant \operatorname{ind}_x \Theta \leqslant 4$$

for each point $x \in \Theta$. A distinctive peculiarity of Θ is that it contains no separating set F of dimension $\dim F \leqslant 1$. This property will turn out to be essential for the axiomatic approach to the definition of dimension (see § 7).

It turns out that an n-dimensional compactum can not only fail to have separating sets of dimension less than n, but may have no closed subsets with dimension between 0 and n. An example of such a compactum which satisfies the first axiom of countability was constructed by Fedorchuk in 1973. Assuming the continuum hypothesis, one can construct perfectly normal compacta without intermediate dimensions. Dropping the first axiom of countability leads to even more striking effects. Assuming the continuum hypothesis, there exist n-dimensional compacta for each $n \geqslant 1$ in which each closed set is either finite or n-dimensional. This example is interesting because, in spite of its dimension-theoretic form, it provides a negative solution of the conjecture in the theory of cardinal invariants that every infinite compactum contains either a minimal $\alpha\mathbb{N}$ or a maximal $\beta\mathbb{N}$ compactification of the space $\mathbb{N}$ of natural numbers.

If the inductive dimension ind of a compactum X is well-defined, then X always has closed subsets F of dimension ind $F = 1$ and, hence, by Theorem 5 of Chapter 1 and the inequalities (3), dim $F = 1$. Therefore, the inductive dimensions are not defined for compacta without subsets of intermediate dimension. In connection with this, it is natural to ask the following question.

Question 1. Suppose that the inductive dimension ind of a compactum X is well defined and that dim $X = n$. Is it true that, for each $k < n$, there is a closed subset F of X with dimension dim $F = k$?

In 1969 Filippov took an essential step towards the solution of the second part of Aleksandrov's question about the relation between the inductive dimensions. He constructed a compactum Φ for which

$$2 = \dim \Phi = \text{ind } \Phi < \text{Ind } \Phi = 3.$$

Lifanov, Pasynkov, and Filippov subsequently published other examples which also give a negative answer to Aleksandrov's question in, among other things, the class of compacta which satisfy the first axiom of countability (see [PFF]). It was proved that the difference between the inductive dimensions can be arbitrarily large. For each $n \geq 1$, Filippov constructed compacta X_n satisfying the first axiom of countability for which

$$\dim X_n = 1, \quad \text{ind } X_n = n, \quad \text{Ind } X_n = 2n - 1.$$

However, the answer to the following is not known.

Question 2. For each triple of natural numbers $1 < m < n$ does there exist a compactum (satisfying the first axiom of countability) for which

$$\dim X = 1, \quad \text{Ind } X = n, \quad \text{ind } X = m?$$

To answer this question affirmatively, it would suffice to construct a compactum Y_n for every $n \geq 2$ such that

$$\dim Y_n = 1, \quad \text{ind } Y_n = 2, \quad \text{Ind } Y_n = n.$$

Closely related to the question of the coincidence of dimensions is the question of whether the sum theorem holds for inductive dimensions. In § 6, we will see that the sum theorem holds for the dimension Ind in the class of perfectly normal spaces. Therefore, for a perfectly normal compactum X,

$$\text{ind } X = \text{Ind } X. \tag{7}$$

However, Filippov has shown that the sum theorem for ind and, consequently, for Ind ceases to hold upon passing from the class of perfectly normal spaces to the class of compacta satisfying the first axiom of countability. Therefore, proposition 5 is all that remains of the sum theorem for inductive dimensions in the class of normal spaces or even compact spaces satisfying the first axiom of countability.

There are examples of perfectly normal compacta in which the dimensions dim and ind do not coincide. To date, such examples have only been found by making additional set-theoretic assumptions such as the existence of a Suslin continuum or the Continuum Hypothesis. Whether there exist "naive" examples of perfectly normal compacta with noncoinciding dimensions dim and ind remains an open question. Alternatively, the following is open.

Question 3. Is the assertion that

$$\dim X = \text{ind } X = \text{Ind } X$$

for every perfectly normal compactum X compatible with axiomatic ZFC set theory?

Questions about the coincidence of the dimensions of compacta have been investigated for other classes of spaces. The following holds.

Proposition 6. *For each compactum X, there exists a dyadic compactum $D(X) \supset X$ with dimensions*

$$\dim D(X) = \dim X, \text{ind } D(X) \leqslant \text{ind } X + 1, \text{Ind } D(X) \leqslant \text{Ind } X + 1.$$

Indeed, we imbed the compactum X in I^τ and let $f: D^\tau \to I^\tau$ be an arbitrary epimorphism. Let $D(X)$ denote the compactum obtained by gluing D^τ to X by the map $f: f^{-1}X \to X$. The equality $\dim D(X) = \dim X$ follows from Theorem 14, Chapter 2. The indicated bounds on the inductive dimensions of $D(X)$ hold because any two disjoint closed subsets of $D(X)$ can be separated by a set in X.

Proposition 6, together with the existence of the aforementioned compacta X_n with fundamental dimensions $1, n, 2n - 1$, imply the existence of dyadic compacta whose fundamental dimensions do not coincide.

An interesting subclass of the class of dyadic compacta is the class $AE(0)$ of compacta or *absolute extensors* in dimension zero; that is, compacta for which the problem of extending partial maps of zero-dimensional compacta is solvable. These compacta are also known as *Dugundji spaces*. Fedorchuk constructed a Dugundji space X for which

$$\dim X = 1, \qquad \text{ind } X = 2.$$

At the same time, the inductive dimensions of Dugundji spaces coincide. They also coincide for the significantly broader class of *openly-generated compacta*; that is, compacta which are limits of sigma-spectra of metrizable compacta and open maps. This class is not contained in the class of dyadic compacta. In order to formulate a more general result along these lines, we will need the following definition.

Definition 1. A space X is called *perfectly kappa-normal* if each canonically closed subset is a G_δ. A space X is called *absolutely kappa-normal* if each closed G_δ set is perfectly kappa-normal.

Theorem 9 (Fedorchuk). *The large and small inductive dimensions of absolutely kappa-normal, completely paracompact spaces coincide.*

Remark 1. It is worth mentioning the large role that completely paracompact spaces play in dimension theory. This class of spaces is much broader than the class of compacta, but shares the main dimensional properties of the latter. In particular, if X is a completely paracompact space, then the inequality dim $X \leqslant$ ind X holds. This inequality was originally obtained by Aleksandrov for compacta.

2.2. Dimension of Subspaces. One of the main questions in dimension theory concerns the monotonicity of dimension; that is, the question of when the dimension does not increase on passing to a subspace. In order that the latter be the case, we either have to impose conditions on the subspace or on the whole space. Thus, we know, that the dimension does not increase under passage to a closed subspace and, consequently, whenever there is a sum theorem, under passage to a F_σ subset. In compacta, F_σ subsets are Lindelöf. A more general monotonicity result involving restrictions on the subspace was obtained by Zolotarev.

Theorem 10. *If A is a completely paracompact subspace of a normal space X, then dim $A \leqslant$ dim X.*

The hypotheses cannot be weakened by replacing complete paracompactness by paracompactness. What is more, there exist zero-dimensional compacta containing metrizable subspaces of positive dimension. For this, we can take Roy's example of an inductively zero-dimensional metric space of Lebesgue dimension 1 (see § 5) and imbed it, by Theorem 6 of Chapter 1, into the Cantor discontinuum D^τ.

One cannot replace the Lebesgue dimension dim in Theorem 10 by the large inductive dimension, even by strengthening the conditions on A. Zolotarev constructed a compactum X of dimension Ind $X = 2$ containing an F_σ subset A of dimension Ind $A = 3$.

A natural class of spaces for which there is a significant theory of dimension is the class of normal spaces. Hence, hereditary normality of a space is a natural restriction to impose for studying monotonicity of dimension. The first hereditarily normal compacta in which monotonicity of the dimensions dim and Ind did not hold were first constructed by Fedorchuk assuming Jensen's combinatorial principle. Compacta were constructed in which the dimensions dim and Ind are not simultaneously monotonic as well as compacta in which dim is monotonic, but Ind is not. Here the dimension increases under passage to a perfectly normal, countably-compact subspace. To construct such examples "naively", that is, without additional set-theoretic assumptions, is impossible. This follows from *Weiss's theorem* which asserts that if Martin's axiom holds and the continuum hypothesis is false, then any perfectly normal countably-compact space is compact.

Assuming the continuum hypothesis, which is weaker than Jensen's principle, there exists a hereditarily normal, zero-dimensional compactum X containing subspaces A_{mn} ($1 \leqslant m \leqslant n$) with dimensions dim $A_{mn} = m$, and Ind $A_{mn} = n$.

Question 4. Do there exist "naive" examples of hereditarily normal compacta in which the dimensions dim and Ind are not monotonic?

§3. The Dimension of Products of Compacta. Cantor Manifolds

3.1. Dimension of Products. In Chapter 2 (Theorem 18), it was proved that the inequality

$$\dim(X \times Y) \leqslant \dim X + \dim Y \tag{8}$$

holds for spaces with a countable base and, in particular, for polyhedra. If X and Y are compact, then any open cover ω of the product $X \times Y$ has an open refinement of the form $u \times v = \{U \times V : U \in u, V \in v\}$ where u and v are open covers of X and Y, respectively. By applying the theorem about ω-mappings (Theorem 7, Chapter 2), it follows from this that the inequality (8) is true for any arbitrary compacta X and Y.

Filippov showed that the logarithmic laws

$$\mathrm{Ind}(X \times Y) \leqslant \mathrm{Ind}\, X + \mathrm{Ind}\, Y \tag{9}$$

and

$$\mathrm{ind}(X \times Y) \leqslant \mathrm{ind}\, X + \mathrm{ind}\, Y \tag{10}$$

for the inductive dimensions of the product of compacta do not hold in general. He constructed compacta X and Y such that $\mathrm{ind}\, X = \mathrm{Ind}\, X = 1$ and $\mathrm{ind}\, Y = \mathrm{Ind}\, Y = 2$, but $\mathrm{ind}(X \times Y) \geqslant 4$.

It is easy to prove that if the finite sum theorem for inductive dimensions holds in the product $X \times Y$ of compacta X and Y, then the inequality (9), which coincides in this case with (10), is satisfied. Pasynkov has proved that inequalities (9) and (10) hold if the finite sum theorem for the inductive dimensions holds in the factors. In this connection, it is worth mentioning that if the finite sum theorem is satisfied for one of the inductive dimensions in a compactum, then $\mathrm{ind}\, X = \mathrm{Ind}\, X$.

The following is interesting.

Question 5. Suppose that the finite sum theorem for inductive dimensions holds in the compacta X and Y. Do the dimensions $\mathrm{ind}(X \times Y)$ and $\mathrm{Ind}(X \times Y)$ coincide?

More generally, one might ask the following.

Question 6. Suppose that the compacta X and Y are such that $\mathrm{ind}\, X = \mathrm{Ind}\, X$ and $\mathrm{ind}\, Y = \mathrm{Ind}\, Y$. Is it true that $\mathrm{ind}(X \times Y) = \mathrm{Ind}(X \times Y)$?

Concerning Pasynkov's theorem about the dimension of products, we remark that Lifanov has obtained the equality

$$\dim \prod_{i=1}^{k} X_i = \mathrm{ind} \prod_{i=1}^{k} X_i = \mathrm{Ind} \prod_{i=1}^{k} X_i = k$$

for compacta X_i of dimension $\mathrm{Ind}\, X_i = 1$.

3.2. Cantor Manifolds

Theorem 11. *Every n-dimensional compactum contains an n-dimensional Cantor manifold.*

This theorem was proved for metric compacta by Hurewicz and Tumarkin and, in general, by Aleksandrov. We sketch the proof. According to Lemma 12 of Chap. 2, which characterizes dimension by extension of maps to the sphere, any n-dimensional compactum contains a closed set A on which there is a continuous map $f_A: A \to S^{n-1}$ which cannot be extended to a continuous map $f: X \to S^{n-1}$. By the Kuratowski-Zorn lemma, there is a closed set $F \subset X$ with the property that f_A cannot be extended to $A \cup F$, but can be extended to any $A \cup F'$ where F' is a proper closed subset of F. We claim that F is an n-dimensional Cantor manifold. Indeed, the inequality $\dim F \geq n$ follows from the aforementioned characterization of dimension by extension of maps to the sphere. Suppose that F is not a Cantor manifold and represent F as a union of two closed sets F_1 and F_2 with $\dim(F_1 \cap F_2) \leq n - 2$. It follows from the definition of F that the map f_A extends to a map $f_1: A \cup F_1 \to S^{n-1}$ and to a map $f_2: A \cup F_2 \to S^{n-1}$. As maps on $A \cup (F_1 \cap F_2)$, the maps f_1 and f_2 are homotopic because $\operatorname{rd} H \leq n - 2$ where $H = \{x: f_1(x) \neq f_2(x)\}$. Therefore, by the mushroom lemma, each extends to $A \cup F$. This contradiction completes the proof.

A maximal n-dimensional Cantor manifold lying in an n-dimensional compactum is called a *dimensional component*. Aleksandrov proved that a dimensional component A of a perfectly normal compactum X is not contained in the union B of the remaining dimensional components. In fact, he also proved that $\dim(A \cap B) \leq n - 2$. Fedorchuk showed that this theorem did not even extend to heredirarily normal compacta satisfying the first axiom of countability.

The union of all dimensional components of an n-dimensional compactum X is called its *internal dimensional kernel* and denoted K_X. The set K_X coincides with the union of all n-dimensional Cantor manifolds in X. Therefore, by Theorem 11, the set $X \backslash K_X$ contains no n-dimensional compacta. Aleksandrov asked about the dimension of the complement of the internal dimensional kernel of a hereditarily normal compactum. Assuming the continuum hypothesis, Fedorchuk constructed hereditarily normal n-dimensional compacta X in which $\dim K_X$ and $\dim(X \backslash K_X)$ can independently take arbitrarily large values, including infinity.

Therefore, the appropriate context for the concept of the internal dimensional kernel is the class of perfectly normal spaces. Here the dimension is monotone under passage to subsets and $\dim(X \backslash K_X) \leq \dim K_X = \dim X$. A.V. Ivanov used the continuum hypothesis to show that the equality $\dim(X \backslash K_X) = \dim X$ can be attained.

If X is a metric compactum, the internal dimensional kernel K_X is contained in Uryson's *inductive kernel*, the latter is the set N_X of all points $x \in X$ for which $\operatorname{ind}_x X = \operatorname{ind} X$. It is clear that $\dim(X \backslash N_X) < \dim X$.

At the same time, beginning with $\dim X = 2$, the following is open.

Question 7. Does there exist a metric compactum X for which

$$\dim(X \setminus K_X) = \dim X?$$

If $\dim X = 1$, let $f: X \to C_X$ denote the quotient map of X onto the zero-dimensional space of connected components of X. Then $f|_{X \setminus K}$ is a homeomorphism and, hence, $\dim(X \setminus K_X) \leqslant 0$.

§4. Locally Finite Covers and Dimension

4.1. The Locally Finite Sum Theorem

Theorem 12 (Katetov, Morita). *Let* $\varphi = \{F_\alpha : \alpha \in \mathscr{A}\}$ *be a locally finite open cover of a normal space* X *by closed sets* F_α *of dimension* $\dim F_\alpha \leqslant n$. *Then* $\dim X \leqslant n$.

As is the case for the countable sum theorem, the proof uses the theorem about the extension of maps to the sphere. After well ordering the index set $\mathscr{A}$, we successively extend the partial maps $f_A: A \to S^n$ by transfinite recursion to elements F_α of the cover φ.

Definition 2. Let X be a normal space. We write locdim $X \leqslant n$ if every point $x \in X$ has a neighbourhood Ox for which $\dim \overline{Ox} \leqslant n$.

From Theorem 12, we immediatey get the following.

Theorem 13 (Dowker, Nagami). *If* X *is paracompact, then* locdim $X = \dim X$.

4.2. Canonical Maps and Nerves of Locally Finite Covers.
The definition, given in §1 of Chapter 2, of an abstract simplicial complex over a fixed set of vertices E nowhere uses the finiteness of the set E and is suitable in general. The definition and main properties of a simplex $T \subset \mathbb{R}^n$ do not depend on the finite-dimensionality of $\mathbb{R}^n$ and, therefore, carry over word for word to a Hilbert space H^τ of arbitrary infinite weight τ.

Definition 3. A complete complex K, whose elements are pairwise disjoint open simplices of a Hilbert space H^τ is called a *triangulation in the Hilbert space* H^τ if the set of vertices of the complex K is a uniformly discrete system of points in H^τ; that is, $\rho(e_\alpha, e^\beta) \geqslant \varepsilon$ for some $\varepsilon > 0$ and all vertices $e_\alpha, e_\beta \in K$.

Proposition 7. *Every complete complex* K *on a set of vertices* E *of infinite cardinality* τ *is isomorphic to a triangulation in a Hilbert space* H^τ.

Proof. Associate to each vertex $e_\alpha \in E$ the unit vector on the αth axis in H^τ; that is, the point X^α all coordinates of which are equal to zero, except the αth which is equal to one. The points $X^{\alpha_0}, \ldots, X^{\alpha_r}$ span a simplex in H^τ if and only if the vertices $e_{\alpha_0}, \ldots, e_{\alpha_r}$ generate the skeleton of a complex K. The set of vertices $\{X^\alpha\}$ of K is discrete in H^τ since different vertices are located at distance $\sqrt{2}$ from one another.

Examples of abstract simplicial complexes are the nerves N_ω of pointwise finite collections ω of subsets of a given set X.

Theorem 14 (Dowker). *For each locally finite open cover ω of a normal space X, there exists a canonical map of X to the underlying polyhedron $\tilde{N}_\omega$ of the nerve N_ω of the cover ω.*

Proof. Choose a partition of unity $\{\mu_\alpha : \alpha \in \mathscr{A}\}$ subordinate to the cover $\omega = \{O_\alpha : \alpha \in \mathscr{A}\}$ and use it to construct a barycentric map $f : X \to N_\omega$ as in the proof of the theorem in Chapter 2. Continuity of f is guaranteed by local finiteness of the cover ω.

Since the cardinality of a locally finite cover of X of weight τ does not exceed τ, Theorem 14 implies the following result.

Corollary 2. *For every cover ω of a paracompact space X of weight τ there exists an ω-map of X to a metric space of weight τ.*

Complement to Theorem 14. *If ω is a star-finite cover, then the canonical map can be assumed essential on the preimage of every simplex of the complex N_ω.*

Indeed, in this case, the star of each vertex contains a simplex of highest dimension and one can apply the process of sweeping out used in the proof of Theorem 5 in Chapter 2.

4.3. Dowker's Theorem

Theorem 15. *For a normal space X the inequality $\dim X \leqslant n$ is equivalent to each of the following conditions:*

a) every star-finite open cover of X has a refinement by an open cover of multiplicity less than or equal to $n + 1$;

b) every locally finite cover of X has a refinement by an open cover of multiplicity less than or equal to $n + 1$.

Proof. Let $\dim X \leqslant n$ and ω be a star-finite open cover of the space X. By the complement to Theorem 14, there exists a canonical map $f : X \to \tilde{N}_\omega$ which is essential on the preimage of every simplex. It follows from the essential map theorem that each simplex of N_ω has a dimension less than or equal to n. Hence the cover by principal stars has multiplicity less than or equal to $n + 1$ and its preimage under f is the desired cover refining ω.

It remains to check that a) implies b). The verification is based on the following two assertions:

1. Every locally finite open cover of a normal space has a refinement which is a locally finite, σ-discrete, open cover;

2. Every countable, locally finite, open cover of a normal space has a refinement which is a star-finite open cover.

In view of Theorem 14, each assertion above reduces to the corresponding assertion about covers of metric spaces.

Corollary 3. *Any open cover of a paracompact space X of dimension $\dim X \leqslant n$ has a refinement by an open cover of multiplicity $\leqslant n + 1$.*

A nice complement to Theorem 15 is the following.

Proposition 8. *If* dim $X \leqslant n$, *then every locally finite open cover of X has a refinement which is a union of no more than $n + 1$ X-discrete collections of open conull-sets.*

According to Theorem 14, it suffices to prove this assertion for a locally finite cover ω of multiplicity $n + 1$ of a metric space X. It is also clear that we need not require openness of the discrete collections. The proof is by induction on n. We pass from $n - 1$ to n as follows. Shrink the cover $\omega = \{U_\alpha : \alpha \in A\}$ to a cover $\omega' = \{V_\alpha : \alpha \in A\}$ such that $\bar{V}_\alpha \subset U_\alpha$. The collection $v_{n+1} = \{V_{\alpha_0} \cap \cdots \cap V_{\alpha_n} : \alpha_0, \ldots, \alpha_n \in A\}$ is discrete. Let V be the body of this system and $Y = X \backslash V$. The cover $\omega' | Y$ has multiplicity less than or equal to n. By the induction assumption, it refines a cover which is a union of discrete collections $v_1, \ldots, v_n$. Then $v = v_1 \cup \cdots \cup v_n \cup v_{n+1}$ is the desired cover.

§5. The Dimension of Metric Spaces

5.1. The Coincidence of the Dimensions dim and Ind. We begin with a characterization due to Vopěnka of the dimension of a metric space.

Theorem 16. *A metric space X has dimension* dim $X \leqslant n$ *if and only if there exists a sequence* $u_1, u_2, \ldots$ *of locally finite open covers of X such that* 1) ord $u_i \leqslant n + 1$; 2) *the collection of the closures of the elements of u_{i+1} is a refinement of u_i;* 3) diam $u_i < 1/i$.

On the one hand, Theorem 16 follows from Corollary 2. Instead, we will prove it together with Theorem 17.

Theorem 17 (Katetov). *For every metrizable space X,*

$$\text{Ind } X = \dim X. \tag{11}$$

Proof. We only need to show that Ind $X \leqslant$ dim X. We proceed by induction on the dimension dim X. If dim $X = 0$ it follows that Ind $X = 0$ since X is normal. To pass from $n - 1$ to n inductively, it suffices to prove that any two disjoint closed subsets A and B of an n-dimensional metric space X can be separated by a set C of dimension dim $C \leqslant n - 1$. We shall prove that there exists a separating set C satisfying the conditions of Theorem 16 with n replaced by $n - 1$. This, together with the equality Ind $X = $ dim X, will also establish Theorem 16 by induction (the case $n = -1$ is obvious). Thus, let $u_1, u_2, \ldots$ be the sequence of covers of X in theorem 16 and let $\rho(A, B) = 1$. Set $K_0 = A$, $M_0 = B$ and, for $i \geqslant 1$, put $K_i = X \backslash H_i$ and $M_i = X \backslash G_i$ where $G_i = \bigcup \{U \in u_i : \bar{U} \cap M_{i-1} = \varnothing\}$ and $H_i = \bigcup \{U \in u_i : \bar{U} \cap K_{i-1} = \varnothing\}$. It is easy to verify that $K_{i-1} \subset X \backslash H_i = \text{Int } K_i$, $M_{i-1} \subset X \backslash G_i = \text{Int } M_i$. Therefore, the sets $K = \bigcup \{K_i : i = 0, 1, 2, \ldots\}$ and $M = \bigcup \{M_i : i = 0, 1, 2, \ldots\}$ are open disjoint neighbourhoods of A and B, respectively. Set $C_i = X \backslash (K_i \cup M_i) = G_i \cap H_i$. The set $C = \bigcap \{C_i : i = 1, 2,$

$\dots\} = X\backslash K \cup M$ separates A and B. The family $v_i = \{U \cap C: U \in u_i$ and $\bar{U} \cap M_{i-1} \neq \varnothing\}$ is an open cover of $C \subset H_i$ and has multiplicity less than or equal to n because each point $x \in C \subset C_i \subset G_i$ belongs to at least one set $U \in u_i$ satisfying the condition $\bar{U} \cap M_{i-1} = \varnothing$. Thus, the sequence $v_1, v_2, \dots$ satisfies the conditions of Theorem 16_{n-1} and this completes the proof of Theorems 16 and 17.

Hence, the dimensions dim and Ind coincide for metric spaces. The small inductive dimension ind coincides with the dimensions dim and Ind in the class of strongly metrizable spaces; that is, spaces having a σ-star-finite base, since these spaces are completely paracompact and Zarelya has shown that if X is completely paracompact, then

$$\dim X \leqslant \text{ind } X. \tag{12}$$

In particular, the dimensions dim, ind, and Ind coincide for strongly paracompact metric spaces. This is not true for an arbitrary metric space. In 1962, Roy constructed an example of an inductively zero-dimensional metric space X having dimension Ind $X = 1$.

Question 8. Does there exist a metric space X for which Ind $X -$ ind $X \geqslant 2$?

5.2. Characterizations of Dimension

Theorem 18. *For any metrizable space X the following conditions are equivalent:*
a) $\dim X \leqslant n$;
b) *X has a σ-locally finite base $\mathscr{B}$ such that* $\dim \text{Bd } U \leqslant n - 1$ *for each* $U \in \mathscr{B}$;
c) *X is a union of $n + 1$ subspaces $X_1, \dots, X_{n+1}$ of dimension* $\dim X_i \leqslant 0$;
d) *X has a σ-locally finite base $\mathscr{B}$ such that* $\text{ord}\{\text{Bd } U: U \in \mathscr{B}\} \leqslant n$.

The implication a) $\Rightarrow$ b) easily follows from the equality $\dim X = \text{Ind } X$. One verifies that b) $\Rightarrow$ a) by showing that disjoint closed subsets A and B of X can be separated by a set C lying in $\bigcup \{\text{Bd } U: U \in \mathscr{B}\}$. Then, by the locally finite sum theorem (Theorem 12) and the countable sum theorem (Theorem 15, Chap. 2), we have $\dim C \leqslant n - 1$ which gives $\dim X \leqslant n$ by Theorem 17.

The implication b) $\Rightarrow$ c) is verified by induction on n. In the induction step, the space $X_{n+1} = X\backslash \bigcup \{\text{Bd } U: U \in \mathscr{B}\}$ is zero-dimensional because the base $\mathscr{B}$ cuts out a base of open-closed sets of it; and the space $\bigcup \{\text{Bd } U: U \in \mathscr{B}\}$ is $(n - 1)$-dimensional by the σ-locally finite sum theorem. The implication c) $\Rightarrow$ a) is the content of Proposition 11, Chapter 1.

To verify that a) $\Rightarrow$ d), we obtain the required base by "thickening" a σ-discrete base. In the process of thickening, we use the following simple assertion.

Proposition. *Suppose that we are given a finite collection $F_1, \dots, F_k$ of closed subsets of a metric space X. Then any disjoint closed subsets A and B of X can be separated by a set C such that* $\text{Ind}(C \cap F_i) < \text{Ind } F_i$ *for all* $i = 1, \dots, k$.

The implication d) $\Rightarrow$ b) $\Rightarrow$ a) is verified by induction on n. For $n = 0$, conditions d) and b) are equivalent. The inductive step from $n - 1$ to n uses the fact that when $U \in \mathscr{B}$ the base $\mathscr{B}$ cuts out a base on Bd U and the collection of sets bounding members of this base has multiplicity less than or equal to $n - 1$.

5.3. Completions, Products, Limits of Inverse Sequences. The following assertion was proved by Tumarkin for spaces with a countable base and by Katetov and Morita in general.

Theorem 19. *Any subspace A of a metric space X with dimension dim $A \leqslant n$ is contained in a G_δ-set A^* of the same dimension* dim $A^* \leqslant n$.

By assertion c) of Theorem 18, it suffices to prove the theorem for zero-dimensional subspaces A. By a modification of assertion (b) of Theorem 18, there exists a σ-locally finite base $\mathscr{B}$ of X such that $A \cap$ Bd $U = \varnothing$ for each $U \in \mathscr{B}$. Then $B = \bigcup \{$Bd $U : U \in \mathscr{B}\}$ is a F_σ set and its complement A^* is zero-dimensional by Theorem 18b).

Since a G_δ subset of a complete metric space is metrizable by a complete metric, Theorem 19 immediately implies the following.

Theorem 20. *Every metrizable space is contained in a topologically complete metrizable space of the same dimension.*

Theorem 21. *If X and Y are metrizable spaces, not both empty, then*

$$\dim(X \times Y) \leqslant \dim X + \dim Y. \tag{13}$$

The proof is by induction on dim X + dim Y using Theorem 18b). Here one needs to note that the product of σ-locally finite bases of X and Y is a σ-locally finite base of $X \times Y$ and

$$\text{Bd}(U \times V) \subset ((\text{Bd } U) \times Y) \cup (X \times \text{Bd } V)$$

Theorem 22. *If a metrizable space X is a limit of an inverse sequence $\{X_i, \pi_j^i\}$ of metric spaces X_i of dimension dim $X_i \leqslant n$ then dim $X \leqslant n$.*

Let the metric ρ_i on X_i be bounded by 1. Consider the metric $\rho = \Sigma(1/2^i)\rho_i$ on the product ΠX_i. Then construct a sequence of covers u_i on X satisfying the conditions of Theorem 16 and such that $u_i = \pi_i^{-1} v_i$.

5.4. The Factorization Theorem and Universal Spaces

Theorem 23 (Pasynkov). *For every continuous map $f : X \to Y$ of metrizable spaces, there exists a metrizable space Z and continuous maps $g : X \to Z$ and $h : Z \to Y$ such that* dim $Z \leqslant$ dim X, $wZ \leqslant wY$, $g(X) = Z$ *and* $f = h \circ g$.

The proof proceeds along the same lines as the proof of Theorem 1. One constructs a countable sequence of locally finite open covers u_i of X with the properties that u_{i+1} is a star refinement of u_i, ord $u_i \leqslant n$, $|u_i| \leqslant wY$ and diam $f(u_i) < 1/i$. The family $\mathscr{U} = \{u_i : i = 1, 2, \ldots\}$ is a basis of a uniformity on X. The separated uniform space Z associated with $\mathscr{U}$ and the associated natural mappings $X \to Z \to Y$ will be those desired. The bound on the dimension of Z is obtained using Theorem 16.

Theorem 24 (Nagata). *For each integer $n \geqslant 0$ and each cardinal number $\tau \geqslant \aleph_0$ there exists a metrizable space $S^{n\tau}$ of weight τ and dimension $\dim S^{n\tau} = n$ which contains a homeomorphic image of any metric space X of weight $wX \leqslant \tau$ and dimension $\dim X \leqslant n$.*

Take $S^{n\tau}$ to be the space Z of Theorem 23 where X is a discrete union of pairwise nonhomeomorphic metric spaces X_α of weight $wX_\alpha \leqslant \tau$ and dimension $\dim X_\alpha \leqslant n$, Y is the generalized Hilbert space H^τ, and $f|_{X_\alpha}$ is inclusion.

5.5. The Metric Dimension

Definition 4. The *metric dimension* $\mu\dim x$ of a metric space X is the least integer n with the property that for any $\varepsilon > 0$ there exists an open ε-cover of X of multiplicity $n + 1$. If there is no such number, then the metric dimension $\mu\dim X$ is defined to be ∞.

Remark 2. It is clear that one can require that the ε-covers in this definition be locally finite. For a completely bounded metric space X (in particular, for a bounded subset of Euclidean space $\mathbb{R}^m$) we can further require that the covers be finite without altering the value of $\mu\dim X$.

It is easy to prove that the inequality $\mu\dim X \leqslant n$ holds for a subset $X \subset \mathbb{R}^m$ if and only if, for every ε, X can be mapped by an ε-shift into an n-dimensional polyhedron $P \subset X$. Therefore, Sitnikov's example in §7, Chap. 2 is an example of a two-dimensional space whose metric dimension is one. It follows from Corollary 3 of §4 that

$$\mu\dim X \leqslant \dim X$$

for any metrizable space X. Katetov extended this to the inequality

$$\mu\dim X \leqslant \dim X \leqslant 2\,\mu\dim X. \tag{14}$$

Sitnikov's example shows that equality can be attained on the right side of (14). For each $k \geqslant 3$, Nemets constructed a complete metric space M in which the difference between $\mu\dim$ and $\dim$ is the largest possible allowed by Katetov's theorem:

$$\dim M = k - 1, \qquad \mu\dim M = [k/2].$$

He also proved a factorization theorem for metric dimensions and established as a corollary that the class of metric spaces X of weight less than or equal to τ and dimension $\mu\dim X \leqslant n$ admits a space which is universal with respect to uniformly continuous inclusions.

Concerning the first part of formula (14), the equality

$$\mu\dim X = \dim X \tag{15}$$

clearly holds for metric compacta. Sitnikov proved that equality (15) also holds for closed subsets of euclidean spaces.

§6. The Dimension of Hereditarily Normal Spaces

The technical basis of this section is the following.

Theorem 25 (The Dowker Addition Theorem). *If a hereditarily normal space X is a countable union of increasing closed subspaces F_i such that $\text{Ind}(F_{i+1} \backslash F_i) \leq n$ then* $\text{Ind } X \leq n$.

Definition 5. We say that a numerical topological invariant I is *hereditarily monotone* in a topological space X if

$$I(A) \leq I(X_0).$$

for any sets $X_0 \subset X$ and $A \subset X_0$. We say I is *hereditary monotone with respect to open subsets* if the same inequality holds whenever A is open in X_0.

Definition 6. An invariant I on a fixed space X *hereditarily satisfies the sum theorem* if, for any subspace $X_0 \subset X$ and any representation $X_0 = \bigcup \{A_i: i = 1, 2, \ldots\}$ as a countable union of closed subsets A_i of X_0, we have

$$I(X_0) = \sup\{I(A_i): i = 1, 2, \ldots\}$$

As an example, recall that the dimension ind is monotone in any topological space and the dimension dim hereditarily satisfies the sum theorem in any hereditarily normal space.

Theorem 25 implies the following theorem.

Theorem 26 (Dowker's recursion theorem). *If the dimension* Ind *is hereditarily monotone with respect to open subsets in a hereditarily normal space X, then it is hereditarily monotone and satisfies the sum theorem.*

The main results of this section are the following.

Theorem 27 (Čech's monotonicity theorem). *If X_0 is any subset of a perfectly normal space X, then* $\text{Ind } X_0 \leq n$.

Theorem 28 (Čech's sum theorem). *If a perfectly normal space X is a countable sum of closed subsets A_i of dimension* $\text{Ind } A_i \leq n$, *then* $\text{Ind } X \leq n$.

Theorems 27 and 28 are proved simultaneously by induction on dimension using Theorem 26. To pass from $n - 1$ to n, it suffices, by Theorem 26, to prove that any open subset U of a perfectly normal space X of dimension $\text{Ind } X = n$ has dimension $\text{Ind } U \leq n$. This is equivalent to showing that if a perfectly normal space X is a union of a sequence of closed sets A_i with $A_i \subset \text{Int } A_{i+1}$ and $\text{Ind } A_i \leq n$, then $\text{Ind } X \leq n$. Let F be closed in X and O_F any neighbourhood of F. We need to find a neighborhood U of F in O_F such that $\text{Ind Bd } U \leq n - 1$. Suppose that $F = \bigcap \{V_i: i = 1, 2, \ldots\}$ where the V_i are open subsets of X such that $\bar{V}_{i+1} \subset V_i$. Since $\text{Ind } A_{i+1} \leq n$, the neighborhood $O_F \cap V_i \cap \text{Int } A_{i+1}$ of $F \cap A_i$ contains a neighborhood U_i with $\text{Ind Bd}_{A_{i+1}} U_i \leq n - 1$. It is easy to see that $\text{Bd}_{A_{i+1}} U_i = \text{Bd } U_i$ and that $U = \bigcup \{U_i: i = 1, 2, \ldots\}$ is such that

Bd $U \subset \bigcup \{\text{Bd } U_i : i = 1, 2, \ldots\}$. Since, by induction, Theorem 28 holds in dimension less than or equal to $n - 1$, we have Ind Bd $U \leqslant n - 1$.

Theorems 27 and 28 have the following corollary.

Corollary 4. *If a perfectly normal space X is a union of two sets A and B, one of which is closed, then* Ind $X = \max\{\text{Ind } A, \text{Ind } B\}$.

For a long time, Čech's question about the monotonicity of the dimension dim in a hereditarily normal space had been unsolved. Dowker asked the analogous question about the monotonicity of the dimension Ind. Filippov solved these problems assuming the existence of Suslin continua. A naive example of a zero-dimensional normal space contaning a set of positive dimension was constructed by E. Pol and R. Pol.

Theorem 28 (and even the finite sum theorem) allows one to easily prove by induction that ind $\beta X \geqslant$ Ind X for perfectly normal spaces X (Smirnov). This, together with the equality Ind $X =$ Ind βX (see Proposition 4), gives the formula

$$\text{ind } \beta X = \text{ind } X = \text{Ind } \beta X. \tag{16}$$

In contrast to the case for the large inductive dimension, the finite sum theorem fails to hold for ind even for metric spaces (van Douwen, Przymusiński). There are metric spaces with ind equal to one which become inductively zero-dimensional upon removing a single point. The countable sum theorem for ind does not hold in metric spaces even in the case when all but one summand consist of a single open-closed point (Prat). All these examples are based on Roy's example (see 5.1).

§7. Axiomatic Characterizations of Dimension

The question of axiomatically characterizing dimension functions dates to the very beginnings of dimension theory. Necessary conditions on any dimension function d, which we will not explicitly restate in what follows, are topological invariance (that is, $dX_1 = dX_2$ for any two homeomorphic spaces X_1 and X_2) and integrality (dX is an integer greater than or equal to -1).

Menger conjectured that the following four axioms characterize the dimension dim in the class $\mathcal{K}$ of all subspaces of Euclidean m-dimensional space $\mathbb{R}^m$:

M1) (the *normalization axiom*) $dI^n = n$, $-1 \leqslant n \leqslant m$, where $I^{-1} = \varnothing$ and $I^0 = \{0\}$;

M2) the *monotonicity axiom*;

M3) the *countable sum axiom*;

M4) the *compactification axiom* – for each space $X \in \mathcal{K}$, there exists a compactification $\tilde{X} \in \mathcal{K}$ such that $dX = d\tilde{X}$.

Menger proved that his conjecture was true for $m \leqslant 2$. For $m \geqslant 3$ the question is still completely open. Recall that for $m \geqslant 4$ it is still an open question as to whether dim satisfies axiom M4 (see §7, Chap. 2). The difficulties associated with axiom M4 can be bypassed by considering the *generalized Menger conjecture* in which the class of subspaces of a fixed Euclidean space $\mathbb{R}^m$ is replaced by the class of subspaces of any Euclidean space. But Shvedov proved that in this class the dimension dim is not the unique function satisfying axioms M1–M4. These axioms are also satisfied by the cohomological dimension $\dim_G$ with respect to any nonzero Abelian group G with a finite number of generators.

Another axiom system was advanced by Aleksandrov:

A1) the *normalization axiom*;

A2) the *finite sum axiom*;

A3) the *Brouwer axiom*: if $dX = n$, there exists a finite open cover ω of the space X such that every image X_ω of X under an ω-map $f: X \to X_\omega$ is such that $dX_\omega \geqslant dX$;

A4) the Poincaré axiom: every space X which is not a single point contains a closed subset A with $dA < dX$ which separates X.

The function dim satisfies axioms A1, A2, A3, and it is easily verified that in any class $\mathcal{K}$ of normal spaces it is the largest function on $\mathcal{K}$ satisfying these axioms. Aleksandrov proved that there is a unique function satisfying axioms A1–A4 in the class of finite-dimensional metrizable compacta. One might conjecture that the dimension dim is the unique function on the class of all finite-dimensional compacta satisfying axioms A1–A4, but it does not satisfy the axiom A4. This follows from Fedorchuk's example of a compactum without intermediate dimensions (see §2).

Shchepin proved that replacing the finite sum axiom in Alexandrov's axiom system by the countable sum axiom gives an axiomatization of the dimension dim in the classes of all separable metric and all metric spaces. In neither case can the countable sum axiom be replaced by the finite sum axiom.

In the other direction, Lokutsievskiĭ proposed that the Poincaré axiom be weakened as follows.

$A_0 4$) Given any space X which is not a single point and any finite cover of X, there exists a closed subset A which separates X and which can be mapped by an ω-map onto a space Y of dimension $dY < dX$.

He proved that dim is the unique function satisfying axioms A1–A3 and $A_0 4$ on the class of compacta. Thus, sharpening Aleksandrov's axioms gives a characterization of the dimension dim in the class of all metric spaces and weakening them gives a characterization in the class of all compacta.

There are other axiomatizations of the dimension functions, including one for Ind (see [PFF]).

Chapter 4
Infinite Dimensional Spaces

§1. Transfinite Dimensions and Countable Dimensional Spaces

1.1. Transfinite Dimensions. The definition of the inductive dimensions ind and Ind for integers $n = -1, 0, 1, 2, \ldots$ given in Chapter 1 extends immediately to all ordinal numbers. Namely, if β is an ordinal number, suppose that we have already defined the class of spaces X which satisfy the inequality Ind $X \leqslant \alpha$ for $\alpha < \beta$. If X is a normal space, we set Ind $X \leqslant \beta$ if any disjoint subsets A and B of X can be separated by a subset C of dimension Ind $C \leqslant \alpha < \beta$.

One defines the *transfinite dimension* ind for the class of regular spaces in a similar manner.

We shall see below that neither dimension ind nor Ind is defined for every space, not even for every metric space.

The following three assertions are easily proved by transfinite induction.

Proposition 1. *If the transfinite dimension* ind X *is defined for a regular space* X, *then* ind A *is defined for any subspace* A *of* X *and*

$$\text{ind } A \leqslant \text{ind } X \tag{1}$$

Proposition 2. *If the transfinite dimension* Ind X *of a normal space* X *is well defined, then* Ind F *is also defined for any closed subspace* F *of* X *and*

$$\text{Ind } F \leqslant \text{Ind } X \tag{2}$$

Proposition 3. *If the transfinite dimension* Ind X *of a normal space* X *is well-defined, then the dimension* ind X *is also defined and*

$$\text{ind } X \leqslant \text{Ind } X. \tag{3}$$

It is easy to verify that the discrete sum $\bigoplus I^n$ of cubes I^n has dimension $\text{int}(\bigoplus I^n) = \omega_0$. At the same time, the large transfinite dimension of $\bigoplus I^n$ is not defined. Indeed, we will prove more.

Let $\mathcal{K}$ be the class of spaces X with a countable base which can be represented as a disjoint sum of open subsets X_n of X of dimension Ind $X_n = n$, $n = 0, 1, 2, \dots$. Then the dimension Ind X of the space X is not well defined.

Suppose that the dimension Ind X is defined for some $X \in \mathcal{K}$. Among the spaces X for which it is defined, take a Y for which Ind $Y = \beta$ is minimal. We have $Y = \bigoplus Y_n$ where Ind $Y_n = n$. In every summand Y_n, there exist disjoint closed sets A_n and B_n which can only be separated by sets of dimension Ind $C_n \geqslant n - 1$. Since Ind $Y = \beta$, the sets $A = \bigcup \{A_n: n = 1, 2, \dots\}$ and $B = \bigcup \{B_n: n = 1, 2, \dots\}$ can be separated by a set C of dimension Ind $C = \alpha < \beta$. Then $C_n = C \cap Y_n$ is such that $n - 1 \leqslant \text{Ind } C_n \leqslant n$. In each C_n choose a closed set Z_{n-1} of dimension Ind $Z_{n-1} = n - 1$. The set $Z = \bigcup \{Z_n: n = 1, 2, \dots\}$ is closed in C and consequently, by (2), we have Ind $Z \leqslant \text{Ind } C = \alpha < \beta$. But $Z \in \mathcal{K}$ and this contradicts the minimality of the dimension Ind Y in the class $\mathcal{K}$.

Definition 1. A collection $\mathcal{B}$ of open subsets of a space X is called a *large base* if, for any closed set F and any neighbourhood OF, there exists $V \in \mathcal{B}$ such that $F \subset V \subset OF$. The smallest cardinality of a large base of a space X is called the *large weight* of X and denoted WX.

It is easy to see that if F is a closed subset of X, then $WF \leqslant WX$. This inequality, together with transfinite induction on the dimension of X, can be used to prove the following.

Theorem 1. *If X is a normal (respectively, regular) space X for which the dimension* Ind X *(respectively* ind X*) is defined and* $WX \leqslant \omega_\alpha$ *(respectively* $wX \leqslant \omega_\alpha$*), then*

$$\text{Ind } X < \omega_{\alpha+1}, \tag{4}$$

(respectively,

$$\text{ind } X < \omega_{\alpha+1}). \tag{5}$$

Corollary 1. *If X is a separable metrizable space for which the dimension* ind X *is defined, then*

$$\text{ind } X < \omega_1. \tag{6}$$

It is obvious that if X is an infinite dimensional compactum, then $wX = WX$. Therefore, Theorem 1 implies the following.

Corollary 2. *If X is a compactum of weight less than or equal to ω_α, then the dimension* Ind X *is well-defined and*

$$\text{Ind } X < \omega_{\alpha+1}.$$

For each transfinite number $\alpha < \omega_1$ Smirnov and Levshenko constructed metrizable compacta S^α and L^α having dimensions

$$\text{Ind } S^\alpha = \alpha, \qquad \text{ind } L^\alpha = \alpha. \tag{7}$$

The compacta S^α are constructed inductively as follows. S^0 is a point, $S^\beta = S^\alpha \times I$ if $\beta = \alpha + 1$ and S^β is the one point compactification of the discrete sum of the compacta S^α, $\alpha < \beta$, if β is a limiting number.

The very existence of compacta L^α, together with Theorem 1, imply that there does not exist a universal space in the class of spaces X with a countable base (or in the class of metrizable compacta X) for which the dimension ind X is well defined. In precisely the same way, the existence of the compacta S^α implies that there is no universal space in the class of metrizable compacta X for which the dimension Ind X is well defined. In §2 it will be proved (Theorem 22) that the large transfinite dimension of a metric space can take only countable values. Despite this, there does not exist a universal space in the class of separable metrizable spaces X for which the dimension Ind X is defined.

Levshenko proved that the compactum S^{ω_0+1} is a union of two homeomorphic compacta Φ_1 and Φ_2 of dimensions

$$\text{ind } \Phi_1 = \text{Ind } \Phi_i = \omega_0, \qquad i = 1, 2.$$

Thus, the finite sum theorem does not hold for transfinite dimensions even in the class of metrizable compacta. An explicit formula for the transfinite dimension of a union of two closed subsets of a hereditarily normal space was given by Pears.

Luxemburg constructed a metrizable compactum Φ with noncoinciding transfinite dimensions

$$\omega_0 + 1 = \text{ind } \Phi < \text{Ind } \Phi = \omega_0 + 2.$$

But in spite of the fact that the large and small transfinite dimensions do not generally coincide, the following is true.

Theorem 2. *If X is a compactum for which* ind X *is defined, then the dimension* Ind X *is also defined.*

This theorem was proved by Levshenko for hereditarily normal spaces and by Fedorchuk in the general case. Thus, the transfinite dimensions ind and Ind of a compactum are well defined simultaneously.

Theorem 3 (Fedorchuk). *If the transfinite dimensions of the compacta X and Y are defined, then so is the transfinite dimension of the product $X \times Y$.*

If X is a completely regular, then

$$\mathrm{ind}(X \times I) \leqslant \mathrm{ind}\, X + 1.$$

Moreover, there exists a metrizable compactum X for which

$$\mathrm{ind}(X \times I) < \mathrm{ind}\, X + 1.$$

Theorem 4 (Luxemburg). *If X is a separable metrizable space for which the dimension* Ind X *is defined, then there exists a metrizable compactification cX such that* Ind $cX =$ Ind X *and* ind $cX =$ ind X.

An arbitrary metrizable space X has a compactification bX with Ind $bX =$ Ind X and ind $bX =$ ind X under the condition ind $X \geqslant \omega_0^2$. It is not possible to replace the existence of the dimension Ind X by the existence of ind X in Theorem 4 because there exists a complete space X with a countable basis and dimension ind $X = \omega_0$ which is not contained in a metrizable compactum with the same dimension ind.

Pasynkov proved a general theorem about compactfications preserving weight and dimension.

Theorem 5. *If X is a normal space with weight $wX = \tau$ and dimension* Ind $X \leqslant \alpha$, *then there exists a compactification bX of X with weight $wbX = \tau$ and dimension* Ind $bX \leqslant \alpha$.

Thus, the class $\mathscr{K}$ of spaces for which the dimension Ind is defined shares many properties of the class of spaces in which Ind is finite. For example, it is closed under passage to closed subspaces and multiplication by compact factors, and there exist compactifications of the same weight. Moreover, the class of hereditarily normal compacta in $\mathscr{K}$ satisfy the countable sum theorem. Namely, we have the following.

Theorem 6 (Fedorchuk). *If a hereditarily normal compactum X is a union of a countably many summands X_i for which the dimension* Ind X_i *is defined, then the dimension* Ind X *is also defined.*

The following is a step towards solving Aleksandrov's problem about the coincidence of the dimension dim and the cohomological dimension $\dim_Z$ (see Chap. 4, § 1).

Theorem 7 (Fedorchuk). *If* Ind X *is defined for a compactum* X, *then*

$$\dim X = \dim_Z X.$$

The proof is by induction on Ind X. It is sufficient to prove that if $\dim_Z X \leqslant n$, then $\dim X \leqslant n + 1$. To prove this, it suffices, in view of the finite sum theorem for dim and the lemmas about shrinking and thickening finite covers, to prove that any disjoint closed subsets A and B of X can be separated by a subset C of dimension $\dim C \leqslant n$. But if Ind $C <$ Ind X, then $\dim C = \dim_Z C \leqslant n$ by the induction assumption, and this concludes the proof.

Question 1. Does every (metrizable) space X for which the dimension Ind X is defined have a compactification bX with weight $wbX = wX$ and dimensions Ind $bX =$ Ind X and ind $bX =$ ind X?

1.2. Spaces of Countable Dimension

Definition 2. A normal space X is called 0-*countable dimensional* (or simply *countable dimensional*) if it is a countable sum of subspaces X_i of dimension $\dim X_i \leqslant 0$.

Theorem 8 (Hurewicz and Wallman). *If a complete metric space* X *is* 0-*countable dimensional, then the dimension* ind X *is defined.*

One cannot replace ind by Ind in this theorem as the space $\bigoplus I^n$ shows.
A partial converse of Theorem 8 is the following.

Theorem 9. *If the dimension* ind X *is well-defined for a strongly metrizable space* X, *then the space* X *is* 0-*countable dimensional.*

This theorem was proved by Hurewicz and Wallman for spaces with a countable basis and by Smirnov in the general case. As will be shown below (Remark 1), the completeness condition in Theorem 8 is essential. The following has not been solved.

Question 2. *Is every metrizable space for which the dimension* ind *is defined* 0-*countable dimensional?*

From Theorems 8 and 9 we get the following.

Corollary 3. *If* X *is a complete separable metric space, the dimension* ind X *is defined if and only if* X *is* 0-*countable dimensional.*

Theorem 10 (Smirnov). *If* Ind X *is defined for a metric space* X *then the space* X *is* 0-*countable dimensional.*

The space $\bigoplus I^n$ shows that the converse to Theorem 10 does not even hold for complete metric spaces. At the same time, Theorems 2 and 8 give the following.

Theorem 11. *The dimension* Ind *of a* 0-*countable dimensional compactum* X *is well defined.*

Theorems 7 and 11 imply the following.

Corollary 4. *If X is a 0-countable dimensional compactum, then*

$$\dim X = \dim_Z X.$$

Theorem 12 (Nagata). *In the class of 0-countable dimensional metric spaces of weight less than or equal to τ, there exists a universal space $S^{\omega_0 \tau}$.*

Remark 1. This theorem shows that the completeness condition in Theorem 8 is essential because the dimension ind of the space $S^{\omega_0 \omega_0}$ is not defined. Indeed, if the dimension ind $S^{\omega_0 \omega_0}$ were defined, then it would be less than ω_1 by Theorem 1. At the same time, by Theorem 9, the space $S^{\omega_0 \omega_0}$ contains all Levshenko compacta L^α, $\alpha < \omega_1$, which contradicts the monotonicity of the dimension ind.

Nagata also proposed the following two characterizations of a 0-countable dimensional metrizable space X:

a) there exists a σ-discrete base of X with the properties that the boundaries of the elements of the base form a pointwise finite collection;

b) for any countable system of disjoint pairs (A_i, B_i) of closed subsets of X, there exists an pointwise finite collection of sets C_i separating A_i and B_i.

1.3. Weakly Countable Dimensional Spaces

Definition 3. A normal space X is said to be *weakly countable dimensional* if it is a countable sum of closed subspaces X_i of finite dimensional spaces $\dim X_i < \infty$.

Theorem 18 of Chap. 3 implies the following.

Proposition 4. *Every weakly countable dimensional metrizable space is 0-countable dimensional.*

This statement cannot be inverted, even for metrizable compacta. On the other hand, Proposition 4 ceases to be true if we leave the class of metrizable spaces: there exist finite-dimensional compacta which are not 0-countable dimensional (see [S]).

The following results pertain to universal weakly countable dimensional spaces.

Theorem 13 (Pasynkov). *In the class of weakly countable dimensional normal spaces of weight less that or equal to τ there exists a universal space Π_τ^ω which is a countable union of n-dimensional compacta Π^n, $n = 1, 2, 3, \ldots$.*

This theorem can be sharpened for spaces with a countable base.

Theorem 14 (Nagata, Smirnov). *The space Q^ω consisting of all points of the Hilbert cube Q for which only finitely many coordinates differ from zero is universal in the class of weakly countable dimensional spaces with a countable base.*

Theorem 15 (Arkhangel'skiĭ). *A universal space exists in the class of weakly countable dimensional metric spaces of weight less than or equal to τ.*

§2. Weakly and Strongly Infinite Dimensional Spaces

2.1. Essential Maps to the Hilbert Cube

Definition 4. A space X is said to be A-weakly infinite dimensional if, for any countably infinite system of disjoint pairs of closed sets (A_i, B_i) there exist sets C_i separating A_i and B_i such that $\bigcap \{C_i : i = 1, 2, \ldots\} = \varnothing$. A space which is not A-weakly infinite dimensional is called A-strongly infinite dimensional.

The definition of an S-weakly infinite dimensional space is obtained from definition 4 if we require that the intersection of finitely many partitions C_i be empty.

It is clear that the property of being A-weakly (respectively, S-weakly) infinite dimensional is inherited by closed subspaces.

Remark 2. The class of A-weakly (respectively, A-strongly) infinite dimensional compacta coincides with the class of S-weakly (respectively, S-strongly) infinite dimensional compacta. Therefore, in the case of compacta, we speak about weakly and strongly infinite dimensional spaces.

Definition 5. A continuous map $f: X \to Q$ is called *essential* if each map $p_n f: X \to I^n$, $n = 1, 2, 3, \ldots$ is essential, where $p_n: Q = I^\infty \to I^n$ is projection onto an n-dimensional face.

Theorem 16 (Levshenko). *A normal space X is S-strongly infinite dimensional if and only if there is an essential map of X to the Hilbert cube.*

Proof. Let (A_i, B_i) be a collection of disjoint pairs of closed subsets of X such that any system of subsets C_i separating A_i and B_i is centered. Let $f_i: X \to I$ be a continuous function equal to 0 on A_i and 1 on B_i. Then the diagonal product $f: X \to Q$ of the maps f_i is an essential map. Conversely, if we let A_i and B_i be preimages of opposite faces Q_{i0} and Q_{i1} of the Hilbert cube Q under an essential map $f: X \to Q$, then any collection of sets separating A_i and B_i will be centered.

A continuous map $g: X \to Q$ is said to be an *admissible variation of a continuous map* $f: X \to Q$ if $f^{-1}Q_{i0} \subset g^{-1}Q_{i0}$ and $f_{-1}Q_{i1} \subset g^{-1}Q_{i1}$ for each $i = 1, 2, \ldots$. The set of points of the Hilbert cube all coordinates of which are different from 0 and 1 is called the *pseudointerior* and will be denoted by p-Int Q.

Definition 6. A continuous map $f: X \to Q$ is called A-*essential* if p-Int $Q \subset gX$ for every admissible variation g of f.

The following is easily proved.

Proposition 5. *Every A-essential map is essential.*

Theorem 17. *A necessary and sufficient condition that a countably paracompact space be A-strongly infinite dimensional is that it possess an A-essential map into the Hilbert cube.*

Theorem 17 implies that the concepts of essential and A-essential maps into Q coincide for compacta. Sklyarenko obtained a stronger result.

Theorem 18. *If $f: X \to Q$ is an essential map of a compactum X to the Hilbert cube Q, then the map $f: f^{-1} I^n \to I^n$ is essential for every face I^n of Q.*

2.2. The Monotonicity, Addition and Sum Theorems. Countable Dimensional and Weakly Infinite Dimensional Spaces

Theorem 19 (Levshenko). *A hereditarily normal space which is a countable union of A-weakly infinite dimensional sets is A-weakly infinite dimensional.*

Corollary 5. *A hereditarily normal (in particular, metrizable) 0-countable dimensional space is A-weakly infinite dimensional.*

Since the identity map of the Hilbert cube is essential, the Hilbert cube is strongly infinite dimensional by Theorem 16. Therefore, Theorem 19 implies the following.

Corollary 6. *The Hilbert cube is not countable dimensional.*

Thus, it follows from Theorem 9 that the dimension ind of the Hilbert cube is not defined.

Theorem 20 (Levshenko). *Each countably paracompact space which is a countable union of closed A-weakly infinite dimensional subsets is A-weakly infinite dimensional.*

Corollary 7. *Each F_σ-subset of an A-weakly infinite dimensional countably paracompact space is A-weakly infinite dimensional.*

Corollary 8. *Each countable dimensional paracompact space is A-weakly infinite dimensional.*

Remark 3. The countable sum theorem and monotonicity with respect to F_σ-subsets fail for S-weakly infinite dimensional spaces. Indeed, it easily follows from Theorem 16 that $\bigoplus I^n$ is S-strongly infinite dimensional. But its Aleksandrov compactification is weakly infinite dimensional by Corollary 5.

However, Sklyarenko proved that the finite sum theorem holds for S-weakly infinite dimensional spaces.

Aleksandrov's question about whether every countable dimensional metrizable compactum is weakly infinite dimensional, and conversely, was open for a long time. Fedorchuk proved that the answer is no in the class of all compacta. He constructed a weakly infinite dimensional compactum satisfying the first axiom of countability which was not only uncountable dimensional, but in which every non zero-dimensional closed subset was uncountable dimensional. Such spaces will be said to be *hereditarily uncountable dimensional*. Here uncountable dimensional means the space is not a countable sum of normal subspaces whose dimension dim is finite. An example of a metrizable weakly infinite dimensional compactum which is uncountable dimensional was constructed by R. Pol.

Question 3. Does there exist a weakly infinite dimensional metrizable compactum which is hereditarily uncountable dimensional?

It follows from Theorem 18 that this question is equivalent to the following.

Question 4. Does there exist a weakly infinite dimensional metrizable compactum which does not contain a subcompactum of positive dimension?

If we waive weak infinite dimensionality in this question, we obtain a well known problem due to Tumarkin. It was solved in 1965 by Henderson who proved that every strongly infinite dimensional metrizable compactum contains infinite dimensional compacta which have no subcompacta with finite positive dimension. Zarelya proved that every strongly infinite dimensional metrizable compactum contains a hereditarily strongly infinite dimensional compactum. Walsh constructed a metrizable compactum in which all subsets (not just closed subsets) of positive dimension are infinite dimensional.

It is easy to see that any compactum which has a well defined transfinite dimension is weakly infinite dimensional. In this connection, the following is interesting.

Question 5. Is every compactum whose transfinite dimension is defined countable dimensional?

The following has been open for a long time.

Question 6. Is the product of two weakly infinite dimensional compacta weakly infinite dimensional?

The answer to this question is not even known in the metrizable case. In the case when one of the factors has a well defined transfinite dimension, the answer to Question 6 is positive (Levshenko).

2.3. The Structure of S-Weakly Infinite Dimensional Spaces

Definition 7. A countable collection of open sets Γ_n, $n = 1, 2, \ldots$, in a space X is called *conveygent* if for any discrete sequence of points x_i, $i = 1, 2, \ldots$, there exists an index after which every point x_i is contained in one of the sets Γ_n. If the system of open sets Γ_n converges, then the set $\Phi = X \setminus \bigcup \{\Gamma_n : n = 1, 2, \ldots\}$ is called the *limit* of the collection.

Theorem 21 (Sklyarenko). *A hereditarily paracompact (in particular, metrizable) space X is S-weakly infinite dimensional if and only if it contains a convergent system of open finite dimensional sets Γ_n with a weakly infinite dimensional compact limit.*

Theorem 22 (Smirnov). *If the dimension* Ind X *of a metric space X is defined, then X is weakly infinite dimensional and* Ind $X < \omega_1$.

The first assertion of Theorem 22 is verified by simply checking that the definition of S-weak infinite dimensionality holds; this is done by induction on Ind. Concerning the second assertion, Theorem 21 states that X contains a compactum Φ which is a limit of a convergent system of finite dimensional sets.

By induction on Ind Φ, one proves that Ind $X \leqslant \beta(\text{Ind } \Phi)$ where $\beta(\alpha)$ is the number defined as follows:

$$\beta(-1) = \omega_0, \qquad \beta(\alpha') = \sup\{\beta(\alpha): \alpha < \alpha'\} + 1.$$

Smirnov also proved that a metric space X is infinite dimensional if and only if X or the complement of some point in X is S-weakly infinite dimensional.

2.4. Compactification of Weakly Infinite Dimensional Spaces. It is easy to verify the following result.

Proposition 6. *A maximal compactification βX of a normal S-weakly infinite dimensional space X is weakly infinite dimensional.*

Sklyarenko proved that an S-weakly infinite dimensional metric space of weight less than or equal to τ possesses a weakly infinite dimensional compactification of weight less than or equal to τ. Pasynkov proved a factorization theorem for mappings of weakly infinite dimensional compacta and used it to obtain a more general results.

Theorem 23. *Every completely regular space X of weight less than or equal to τ which has a weakly infinite dimensional maximal compactification βX (in particular, every S-weakly infinite dimensional normal space) has a weakly infinite dimensional compactification bX of weight less than or equal to τ.*

Remark 4. It is not possible to replace S-weak infinite dimensionality by A-weak infinite dimensionality in the formulation of Theorem 23. Indeed, the space Q^ω, being countable dimensional, is A-weakly infinite dimensional by Corollary 5. However, Sklyarenko proved that every compactification of Q^ω is strongly infinite dimensional.

2.5. Infinite Dimensional Cantor Manifolds

Definition 8. A compactum X is said to be an *infinite dimensional Cantor manifold* if it cannot be separated by a weakly infinite dimensional closed subset.

The analogue of Theorem 11 in Chap. 3 is the following.

Theorem 24 (E.G. Sklyarenko). *Every strongly infinite dimensional compactum contains an infinite dimensional Cantor manifold.*

Chapter 5
Cohomological Dimension

§ 1. The Definition of Cohomological Dimension.
The Inequality $\dim_G X \leqslant \dim_z X \leqslant \dim X$ and
the Equality $\dim_z X = \dim X$ for Finite Dimensional Compacta.
P.S. Aleksandrov's Problem

1.1. Properties of Cohomology Groups with Compact Support. If X is a locally compact space, G an abelian group, and n a nonnegative integer, then $H_c^n(X; G)$ will denote the nth cohomology group with compact support and coefficients in G.

If F is a closed subset of a compactum X, there exists a natural isomorphism between the groups $H_c^n(X \setminus F; G)$ and the spectral Aleksandrov-Čech cohomology groups $H_c^n(X \setminus F; G)$.

H_c^n is a contravariant functor from the category of locally compact spaces and perfect maps to the category of abelian groups. For a perfect map $f: X \to Y$ the induced homomorphism

$$H_c^n(f): H_c^n(Y) \to H_c^n(X)$$

will be denoted by f^* for short.

Henceforth, X will denote a locally compact space, F a closed subset, and U an open subset. If $f: F \to X$ is an inclusion, the homomorphism $f^*: H_c^n(X; G) \to H(F; G)$ will be denoted as $j_{X,F}$. We let

$$i_{U,X}: H_c^n(U; G) \to X; G)$$

denote the homomorphism dual to the inclusion of pairs

$$(X, \varnothing) \to (X, X \setminus U).$$

Let

$$\partial: H_c^n(F; G_0) \to H_c^{n+1}(X \setminus F; G)$$

denote the coboundary operator.

The homomorphisms ∂, i, j are related to one another by the exact sequence of the pair (X, F)

$$\ldots \xrightarrow{j} H_c^{n-1}(F; G) \xrightarrow{\partial} H_c^n(X \setminus F; G) \xrightarrow{i} H_c^n((X; G) \xrightarrow{j} H_c^n(F; G) \to \ldots \qquad (1)$$

The *universal coefficient formula*

$$H_c^n(X; G_1 \otimes G_2) \cong (H_c^n(X; g_1) \otimes G_2) \oplus \text{Tor}(H_c^{n+1}(X; G_1), G_2) \qquad (2)$$

for $\text{Tor}(G_1, G_2) = 0$ allows us to reduce the computation of cohomology groups with coefficients in an arbitrary group to the computation of the cohomology with coefficients in the group Z of integers.

The *Kunneth formula*

$$H_c^n(X \times Y; G_1 \otimes G_2) \cong \sum_{i+j=n} H_c^i(X; G_1) \otimes H_c^j(Y; G)$$

$$\otimes \sum_{i+j=n+1} \mathrm{Tor}(H_c^i(X; G_1), H_c^j(Y; G)) \qquad (3)$$

allows us to compute the cohomology of product of spaces. Sometimes the following formula is more convenient

$$H_c^n((X \times Y; G) \cong \sum_{i+j=n} H_c^i(X; H_c^j(Y; G)). \qquad (4)$$

If $S = \{X_\alpha, \pi_\beta^\alpha; \alpha \in A\}$ is an inverse spectrum with perfect projections and $X = \varprojlim S$, then

$$H_c^n(X; G) \cong \varinjlim \{H_c^n(X_\alpha; G), (\pi_\beta^\alpha)^*; \alpha \in A\}. \qquad (5)$$

If $\{U_\alpha : \alpha \in \mathscr{A}\}$ is a collection of open sets of X directed above by an inclusion and $X = \bigcup\{U_\alpha : \alpha \in \mathscr{A}\}$, then

$$H_c^n(X; G) \cong \varinjlim \{H_c^n(U_\alpha; G), i_{U_\alpha U_\beta} : \alpha \in \mathscr{A}\}. \qquad (6)$$

1.2. The Definition and Simplest Properties of Cohomological Dimension. The *homological dimension of a space* X with respect to a nonzero coefficient group G is the largest integer n such that there exists a closed set $A \subset X$ for which the Aleksandrov-Čech homology group $H_n(X, A; G)$ is different from zero. The study of homological dimension is hampered by the fact that the exactness axiom does not hold for Alexandrov-Čech homology. Therefore, by the homological theory of dimension, one usually understands the more workable theory of cohomological dimension. For metrizable compacta, where Pontryagin duality holds between the groups $H_n(X, A; G)$ and $H^n(X, A; G^*)$ the homological approach with coefficients in a compact group G is equivalent to the cohomological approach with coefficients in the dual discrete group G^*.

Definition 1. The *cohomological dimension* $\dim_G X$ of a locally compact space X with respect to an abelian coefficient group G is the greatest integer n for which there exists a locally compact set $A \subset X$ such that $H_c^n(A; G) \neq 0$.

It follows immediately from the definition that the cohomological dimension is hereditary. Namely, the following result holds.

Proposition 1. *For any locally compact set $A \subset X$ we have*

$$\dim_G A \leqslant \dim_G X \qquad (7)$$

From the remark at the outset of Section 1.1 regarding the coincidence of $H_c^n(X \setminus F; G)$ and $\check{H}_c^n(X, F; G)$ for compact groups, we obtain the following result.

Proposition 2. *For any compactum X, we have*

$$\dim_G X \leqslant \dim X. \qquad (8)$$

The universal coefficient formula (2) implies the following result.

Proposition 3. *If* X *is locally compact, then*

$$\dim_G X \leqslant \dim_Z X. \tag{9}$$

Theorem 1. *The following conditions are equivalent:*
1) $\dim_G X \leqslant n$;
2) $H_c^{n+1}(U; G) = 0$ *for every open subset* U *of* X *with compact closure*;
3) *the homomorphism* $j_{A,B}: H_c^n(A; G) \to H_c^n(B; G)$ *is an epimorphism for any pair* $B \subset A$ *of compact sets of the space* X.

The implication 1) $\Rightarrow$ 2) is obvious. We verify the implication 2) $\Rightarrow$ 3). It follows from property (6) that condition 2) is satisfied for all open sets $U \subset X$. Since the sequence

$$H_c^n(X; G) \xrightarrow{j_{X,B}} H_c^n(B; G) \xrightarrow{\partial} H_c^{n+1}(X \setminus B; G)$$

is exact and $H_c^{n+1}(X \setminus B; G) = 0$, the map $j_{X,B}$ is an epimorphism. But, $j_{X,B} = j_{A,B} \circ j_{X,A}$ from which it follows that $j_{A,B}$ is an epimorphism.

Instead of the implication 3) $\Rightarrow$ 1), we verify the logically equivalent implication 3) $\Rightarrow H_c^{n+1}(B; G) = 0$ for any compact $B \subset X$. Suppose that $0 \neq e \in H_c^{n+1}(B; G)$. Then the compact set B contains a closed set F satisfying the conidition $j_{B,F}(e) \neq 0$ and minimal with respect to this property. Since $n + 1 \geqslant 1 > 0$, the set F can be written as a union of two proper closed subsets F_1 and F_2. Let

$$H_c^n(F_1; G) \otimes H_c^n((F_2; G) \xrightarrow{\psi} H_c^n(F_1 \cap F_2; G)$$

$$\xrightarrow{\Delta} H_c^{n+1}(F; G) \xrightarrow{\varphi} H_c^{n+1}(F_1; G) \otimes H_c^{n-1}(F_2; G)$$

be the Maier-Vietoris exact sequence (the addition sequence of a triad) where $\varphi = (j_{B,F}, j_{B,F})$, $\psi(a, b) = j_{F_1, F_1 \cap F_2}(a) - j_{E_2, F_1 \cap F_2}(b)$ and $\Delta = i_{F \setminus F_1 \cap F_2, F} \circ \partial)$. By condition 3), the homomorphism ψ is an epimorphism. But then φ is a monomorphism and $\varphi(j_{B,F}(e)) \neq 0$ which contradicts the minimality of F.

From condition 3) of Theorem 1 we obtain the following.

Proposition 4. *The cohomological dimension of any locally compact space is equal to the maximum of the cohomological dimensions of its compact subsets.*

1.3. The Hopf Theorem and Aleksandrov's Identity $\dim_Z X = \dim X$ for Finite Dimensional Compacta

Hopf's Theorem. *Let* X *be a compactum of dimension less than or equal to* $n + 1$ *and let* A *be a closed subset. A necessary and sufficient condition that a map* $f: A \to S^n$ *extend to* X *is*

$$f^*(H_c^n(S_c^n; Z)) \subset j_{X,A}(H_c^n(X; Z)). \tag{10}$$

Necessity follows because $f^* = j_{X,A} \circ \bar{f}^*$ for any extension $\bar{f}: X \to S^n$ of f. Using barycentric coordinates, the proof of sufficiency reduces to the proof of the analogous assertion for an $(n + 1)$-dimensional polyhedron X and a sub-polyhedron A. The heart of the matter is the case $(X, A) \approx (B^{n+1}, S^n)$. Here we must show if $f: S^n \to S^n$ is such that the homomorphism $f^*: H^n(S^n) \to H^n(S^n)$ is zero, then f is an inessential map. But this follows because the homomorphism $f_w\colon H_n(S^n) \to H_n(S^n)$ is also zero and the Hurewicz homomorphism

$$\gamma_n\colon \pi_n(S^n) \to H_n(S^n)$$

is an isomorphism.

Theorem 2. *For a finite-dimensional compactum* X,

$$\dim_{\mathbf{Z}} X = \dim X. \tag{11}$$

Proof. By Proposition 2, it is sufficient to establish that $\dim X \leqslant \dim_{\mathbf{Z}} X$. For this, it suffices to verify that if $\dim X \leqslant n + 1$ and $\dim_{\mathbf{Z}} X \leqslant n$, then $\dim X \leqslant n$. For this in turn, according to the characterization of dimension by extensions of mappings to the sphere (Theorem 12' of Chap. 2), we need only extend $f\colon A \to S^n$ to X. But by condition 3) of Theorem 1 the homomorphism $j_{X,A}$ is an epimorphism. Therefore, it is possible to apply the Hopf theorem and this completes the proof.

Remark 1. The equality (11) holds for any finite-dimensional, locally compact and paracompact space X. This follows from Theorem 2 upon noting that X is a discrete union of σ-compact sets in this case and $\dim X = \max\{\dim F\colon F \subset X$ is compact$\}$ by the sum theorem.

1.4. Aleksandrov's Problem. Aleksandrov's general problem is formulated as follows. Can the cohomological dimension of a compactum X with respect to some coefficient group be finite, but $\dim X = \infty$?

A particular case of the general problem is *Aleksandrov's problem*. Is (11) true for arbitrary compacta?

Remark 2. Forumulating Aleksandrov's general problem in the class of metrizable compacta is, in fact, equivalent to the general problem in the class of all compacta. Indeed, the following factorization theorem holds.

Theorem 3. ([Z]) *Let* $f\colon X \to Y$ *be a map of compacta and* $\dim_G X \leqslant n$. *Then there exists a metrizable compactum* Z *with* $\dim_G Z \leqslant n$ *and maps* $g\colon X \to Z$ *and* $h\colon Z \to Y$ *such that* $f = h \circ g$.

Assume that there exists an infinite dimensional compactum X for which $\dim_G X \leqslant n$. According to Theorem 12 of Chap. 2, there is an essential map $f_m\colon X \to I^m$ for each natural number m. Applying Theorem 3, we obtain a metrizable compactum Z_m with dimension $\dim_G Z_m \leqslant n$ and maps $g_m\colon X \to Z_m$ and $h_m\colon Z_m \to I^m$ such that $f = h_m \circ g_m$. But it is clear that the left factor of an essential map is an essential map. Therefore, $\dim Z_m \geqslant m$ by the Theorem 12 of Chap. 2 again. Let Z be the Alexandrov compactification of the discrete sum $\bigoplus Z_m$. Then $\dim Z = \infty$. In addition, $\dim_G Z \leqslant n$ (the latter follows immediately from both the sum theorem (see §2) and Theorem 1).

Remark 3. If the cohomological dimension of a compactum X is equal to zero for a single coefficient group G, then X is totally disconnected and $\dim X = 0$.

Indeed, if x and y were different points belonging to the same connected component A of X, then the homomorphism

$$j: H_c^0(A; G) \to H_c^0(\{x, y\}; G)$$

would be an epimorphism, contradicting Theorem 1.

Let $K(G, n)$ be an Eilenberg-Maclane complex; that is, a CW-complex for which

$$\pi_k(K(G, n)) = \begin{cases} 0 & \text{if } k \neq n, \\ G & \text{if } k = n. \end{cases}$$

The following is a standard result in homotopy theory.

Theorem 4. *A necessary and sufficient condition that every continuous map from a closed subset A of a compactum X to $K(G, n)$ extend to X is that the homomorphism $j_{X,A}: H_c^n((X; G) \to H_c^n(A; G)$ be an epimorphism.*

This, together with Theorem 1, gives another characterization of cohomological dimension.

Theorem 5. *If X is compact, $\dim_G X \leqslant n$ if and only if all continuous maps of closed subsets $A \subset X$ to the Eilenberg-Maclane complex $K(G, n)$ extend to X.*

Since $K(\mathbf{Z}, 1) = S^1$, Theorem 5 and Theorem 12' of Chap. 2 imply the following corollary.

Corollary 1. *If X is a compactum for which $\dim_{\mathbf{Z}} X \leqslant 1$, then $\dim X \leqslant 1$.*

As the manuscript of this article was being prepared for the printer, Aleksandrov's problem was solved by Dranishnikov. For any $n \geqslant 3$ he constructed a strongly infinite dimensional compactum X with $\dim_{\mathbf{Z}} X = n$.

§2. Aleksandrov's Obstruction Theorem and its Corollaries

Definition 2. We say that a locally compact space X is an *r-dimensional obstruction* at a point $x \in X$ with respect to a coefficient group G if there exists a neighbourhood U of x such that, for any smaller neighbourhood V of x, the image of the homomorphism $j_{V,U}: H_c^r(V; G) \to H_c^r(U; G)$ is different from zero.

Theorem 6 ([K1]). *Let X be a locally compact space with dimension $\dim_G X = r$. Then there exists a locally closed set $A \subset X$ with compact closure such that*
1) $\dim_G A = r$;
2) *X is an r-dimensional obstruction at each point $x \in A$ with respect to the coefficient group G;*
3) *for any open subset W of A with $\overline{W}^A \neq A$, the homomorphism*

$$j: H_c^{r-1}(\overline{W}^A; G) \to H_c^{r-1}(\mathrm{Bd}_A W; G)$$

is not an epimorphism.

Proof. By theorem 1 there exists an open set $U \subset X$ with compact closure such that $H_c^r(U; G) \neq 0$. Suppose that $0 \neq e \in H_c^r(U; G)$ and let A be a closed subset of U such that $j_{U,A}(e) \neq 0$. Let A be minimal with respect to this property. Then $\dim_G A = r$.

Suppose now that $x \in A$ and V is a neighbourhood of x contained in U. The bottom row in the commutative diagram

$$
\begin{array}{ccc}
H_c^r(V; G) & \xrightarrow{\ \ i\ \ } & H_c^r(U; G) \\
\big\downarrow{\scriptstyle j} & & \big\downarrow{\scriptstyle j} \\
H_c^r(V \cap A; G) & \xrightarrow{\ \ \gamma\ \ } & H_c^r(A; G) \xrightarrow{\ \ j\ \ } H_c^r(A \backslash V; G)
\end{array}
$$

is exact, as it is a piece of the exact sequence of the pair $(A, A \backslash V)$. It follows from $H_c^{r-1}(V; G) = 0$ and the exactness of the sequence of the pair $(V, V \cap A)$ that $j_{V, V \cap A}$ is an epimorphism. Finally, since A is minimal, it follows that $j_{A, A \backslash V}(j_{U, A}(e)) = 0$. Therefore, there exists an element $\tilde{e}_1 \in H_c^r(V, G)$ such that $i_{V \cap A, A}(j_{V, V \cap A}(e_1)) = j_{U, A}(e) \neq 0$. But then $i_{V, U}(e_1) \neq 0$. Condition 2) is verified.

Condition 3) follows easily from the minimality of the set A and the exactness of the additive sequence of the triad $(A, \overline{W}^A, A \backslash W)$. Theorem 6 is proved.

Recall that for every n-dimensional orientable manifold M without boundary (not necessarily compact) the Poincaré duality isomorphism

$$D_M: H_c^k(M; G) \to H_{n-k}^c(M; G),$$

where $H_{n-k}^c(M; G)$ is the singular homology group, is well defined. The isomorphism is natural in the sense that, for each open subset U of M, the diagram

$$
\begin{array}{ccc}
H_c^k(U; G) & \xrightarrow{\ \ D_U\ \ } & H_{n-k}^c(U; G) \\
\big\downarrow{\scriptstyle i} & & \big\downarrow{\scriptstyle i_*} \\
H_c^k(M; G) & \xrightarrow{\ \ D_M\ \ } & H_{n-k}^c(M; G)
\end{array}
$$

commutes where i_* is the homomorphism of homology groups induced by inclusion.

Theorem 6' below follows from Theorem 6 by Poincaré duality. It is the original formulation of Aleksandrov's obstruction theorem and characterizes the homological dimension of a closed subset of Euclidean space by the metric and homological properties of its complement.

Theorem 6'. *Let X be a closed subset of n-dimensional Euclidean space of dimension $\dim_G X = r$. Then there exists a point $x \in X$ and an open ball U^n centered at x such that every smaller ball U_1^n centered at x contains an $(n - r - 1)$-dimensional cycle which is not homotopic to zero in $U^n \backslash X$.*

Moreover, for $k < n - r - 1$, any k-dimensional cycle of an arbitrary ball U^n lying in $U^n \backslash X$ is homologous to zero in $U^n \backslash X$. For $k = 0$, one considers the reduced homology group $H_0^c(U^n \backslash X; G)$.

For $G = \mathbf{Z}$ this theorem solved the problem posed by Uryson of *characterizing closed r-dimensional sets r $\leqslant$ n* in terms of their disposition in $\mathbb{R}^n$. Uryson's general problem (that of characterizing any, not necessarily closed, *r*-dimensional subsets) was solved by Sitnikov (see § 5).

If X is closed in $\mathbb{R}^n$ and U^n is an open ball, then $H_c^n(X \cap U^n; G) \neq 0$ if and only if $U^n \subset X$. Consequently, X generates an *n*-dimensional obstruction at the point x if and only if X contains a ball centered at x. Therefore, Theorem 6' implies the following.

Proposition 5. *A closed subset $X \subset \mathbb{R}^n$ has $\dim_G X = n$ if and only if X has interior points.*

This is complemented by the following assertion.

Proposition 6. *A closed subset $X \subset \mathbb{R}^n$ has $\dim_G X = n - 1$ if and only if X is nowhere dense and separates some ball.*

Thus, for closed subsets of $\mathbb{R}^n$ of dimension n (or $n - 1$), the cohomological dimensions coincide for all coefficient groups.

Corollary 2. *If X is a closed subset of three-dimensional Euclidean space, then $\dim_G X = \dim X$ for any coefficient group G.*

Theorem 7. *If each point of a locally compact space X has a neighbourhood whose cohomological dimension $\dim_G$ does not exceed n, then $\dim_G X \leqslant n$.*

Theorem 8 (The countable sum theorem). *If a locally compact space X is a union of a locally countable family $\{X_\alpha\}$ of locally compact subsets, then*

$$\dim_G X = \max_\alpha \dim_G X_\alpha.$$

Proof. Taking into account Proposition 1, it suffices to prove that $\dim_G X \leqslant \max \dim_G X_\alpha$. Upon applying Theorem 7, we may assume that the collection $\{X_\alpha\}$ is countable. Let $A \subset X$ be the set provided by Theorem 6. One of the summands $X_\alpha \cap A$ contains an open subset of A. Condition 3) of Theorem 6) implies that $\dim_G X_\alpha \cap A = \dim_G A$. Hence, $\dim_G X_\alpha \geqslant \dim_G X_\alpha \cap A = \dim_G A = \dim_G X$.

§ 3. Relations Between Cohomological Dimensions with Respect to Different Coefficient Groups

Let $\mathbf{Z}_p$ be the cyclic group of order p; $\mathbb{R}_p$ the group of rational numbers having denominator relatively prime to p; $\mathbb{Q}_p \cong \mathbb{Q}/\mathbb{R}_p$ the *p*-primary quasicyclic group, and p a prime number. Then there exist exact sequences

$$0 \to \mathbf{Z}_p \to \mathbb{Q}_p \overset{P}{\to} \mathbb{Q}_p \to 0$$

$$0 \to \mathbb{R}_p \overset{P}{\to} \mathbb{R}_p \to \mathbf{Z}_p \to 0$$

$$0 \to \mathbb{R}_p \to \mathbb{Q} \to \mathbb{Q}_p \to O$$

$$0 \to \mathbb{Z}_{p^k} \to \mathbb{Z}_{p^{k+1}} \to \mathbb{Z}_p \to O$$

Moreover the group $\mathbb{Q}_p$ is the limit of the direct spectrum of the groups $\mathbb{Z}_{p^k}$.

On the other hand, for every exact sequence of groups

$$0 \to G_1 \to G \to G_2 \to 0$$

there exists an exact sequence

$$\cdots \to H_c^n(X; G_1) \to H_c^n(X; G) \to H_c^n(X; G_2) \xrightarrow{\beta} H_c^{n+1}(X; G_1) \to$$

where the map β is the *Bokshteĭn coboundary operator*. Moreover, $\beta = 0$ if G_1 is a servant subgroup of G (that is, for any $g \in G_1$, the equation $nx = g$ is solvable in G only if it is solvable in the subgroup G_1).

In addition, if G is a limit of a direct spectrum of groups $\{G_\alpha\}$, then $H_c^n((X; G) \cong \varinjlim H_c^n(X; G_\alpha)$.

The following is an easy consequence of the aforementioned properties.

Bokshteĭn's Inequalities. *If X is a locally compact space, then*

$$\dim_{\mathbb{Q}_p} X \leqslant \dim_{\mathbb{Z}_p} X \tag{12}$$

$$\dim_{\mathbb{Z}_p} X \leqslant \dim_{\mathbb{Q}_p} X + 1 \tag{13}$$

$$\dim_{\mathbb{Q}} X \leqslant \dim_{\mathbb{R}_p} X \tag{14}$$

$$\dim_{\mathbb{Z}_p} X \leqslant \dim_{\mathbb{R}_p} X \tag{15}$$

$$\dim_{\mathbb{Q}_p} X \leqslant \max\{\dim_{\mathbb{Q}} X, \dim_{\mathbb{R}_p} X - 1\} \tag{16}$$

$$\dim_{\mathbb{R}_p} X \leqslant \max\{\dim_{\mathbb{Q}} X, \dim_{\mathbb{Q}_p} X + 1\} \tag{17}$$

These inequalities play a large role in the homological theory of dimension because Bokshteĭn found a way to compute the cohomological dimension of a space with respect to a coefficient group G if the cohomological dimensions with respect to the countable collection σ of groups $\mathbb{Q}_p$, $\mathbb{Z}_p$, $\mathbb{R}_p$, $\mathbb{Q}_p$, where p runs over the prime numbers, are known.

If G is an abelian group, define a set $\sigma(G) \subset \sigma$ as follows. A group in σ belongs to $\sigma(G)$ if and only if it satisfies the following conditions:

a) for $\mathbb{Q}$: G contains an element of infinite order;
b) for $\mathbb{Z}_p$: G contains an element of order p^k not divisible by p.
c) for $\mathbb{Q}_p$: G contains an element of order p;
d) for $\mathbb{R}_p$: G contains an element g with the property that $p^n g$ is not divisible by p^{n+1} for any integer $n \geqslant 0$.

Theorem 9. *If X is a locally compact space, then*

$$\dim_G X = \max\{\dim_H X : H \in \sigma(G)\}. \tag{18}$$

This theorem immediately implies the following result.

Corollary 3. *For any locally compact space X,*

$$\dim_{\mathbb{Z}} X = \max\{\dim_{\mathbb{R}_p} X; p \text{ prime}\}. \tag{19}$$

§4. The Cohomological Dimension of a Product of Spaces. Pontryagin's Compacta

4.1. The Dimension of Products. From Theorem 6 about obstructions, we obtain the identity

$$\dim_G(X \times Y) = \max\{n: H^n_c((U \times V; G) \neq \varnothing\} \tag{20}$$

where $U \subset X$ and $V \subset Y$ are open sets.

Therefore, using the Kunneth formula (3) and Bockstein's inequalities, we easily deduce the following result.

Theorem 10. *If X and Y are any locally compact spaces, then*

$$\dim_{\mathbb{Z}_p}(X \times Y) = \dim_{\mathbb{Z}_p} X + \dim_{\mathbb{Z}_p} Y; \tag{21}$$

$$\dim_{\mathbb{Q}}(X \times Y) = \dim_{\mathbb{Q}} X + \dim_{\mathbb{Q}} Y; \tag{22}$$

$$\dim_{\mathbb{Q}_p}(X \times Y) = \max\{\dim_{\mathbb{Q}_p} X + \dim_{\mathbb{Q}_p} Y, \dim_{\mathbb{Z}_p}(X \times Y) - 1\}, \tag{23}$$

if $\dim_{\mathbb{R}_p} = \dim_{\mathbb{Q}_p}$ for at least one on the spaces X and Y, then

$$\dim_{\mathbb{R}_p}(X \times Y) = \dim_{\mathbb{R}_p} X + \dim_{\mathbb{R}_p} Y; \tag{24}$$

if $\dim_{\mathbb{Q}_p} X < \dim_{\mathbb{R}_p} X$ and $\dim_{\mathbb{Q}_p} Y < \dim_{\mathbb{R}_p} Y$, then

$$\dim_{\mathbb{R}_p}(X \times Y) = \max\{\dim_{\mathbb{Q}_p}(X \times Y) + 1, \dim_{\mathbb{Q}}(X \times Y)\}. \tag{25}$$

4.2. Pontryagin's Compacta. Let $K = \{z \in \mathbb{R}^2 = \mathbb{C}: 1 \leqslant |z| \leqslant 2\}$ be an annulus in the plane bounded by two circles. Denote the smaller circle by S^1_1 and the larger by S^1_2. Consider the p-fold cover $f_p: S^1_1 \to S^1_1$ of S^1_1 given by $f_p(z) = z^p$. Let M_p denote the quotient space of K under the equivalence relation $\sim$ whose nonsingleton classes lie in S^1_1 and are such that $z_1 \sim z_2 \Leftrightarrow f_p(z_1) = f_p(z_2)$. The space M_p is a polyhedron called the *Moebius band modulo p*. It is clear that M_2 is the usual Moebius band.

Now let L be a two-dimensional polyhedron situated in $\mathbb{R}^5$. For each two-dimensional simplex T^2 of a triangulation of L, we remove a small disk $D^2 \subset T^2$ and sew in a copy of M_p by gluing Bd D^2 to the boundary S^1_2. We do this so that the Moebius band we attached has no self-intersections and no intersections with the rest of the polyhedron L in $\mathbb{R}^5$. The result of attaching the copies of M_p is a new polyhedron denoted $M_p(L)$. It is clear that $M_p(\bar{T}^2)$ is homeomorphic to M_p.

Set $L = L_{0,p}$, $M_p(L) = L_{1,p}$ and define a map $f^1_0: L_{1,p} \to L_{0,p}$ which, for each simplex $\bar{T}^2_r$ of the triangulation of $L_{0,p}$, maps $M_p(\bar{T}^2_r)$ to $\bar{T}^2_r$ so that $f^1_0|$Bd $\bar{T}^2_r = id$, $f^1_0(M_p)$ is the center vr of the simplex $\bar{T}^2_r$ and $M_p(\bar{T}^2_r)\backslash M_p = \bar{T}^2_r \backslash D^2$ is mapped homeomorphically to $\bar{T}^2_r \backslash \{vr\}$.

Now take a sufficiently find triangulation of the polyhedron $L_{1,p}$ and construct, as above, a polyhedron $L_{2,p}$ and a map $f^2_1: L_{2,p} \to L_{1,p}$ and so on. Set $L_p = \varprojlim \{L_{k,p}, f^k_{k-1}\}$ where $L_{0,p} = \bar{T}^2$.

Let p and q be different prime numbers. We shall prove that

$$\dim_{\mathbf{Z}_p}(L_p) = \dim_{\mathbf{R}_p}(L_p) = \dim L_p = 2 \qquad (26)$$

and

$$\dim_G L_p = 1 \qquad (27)$$

if G is any of the following groups

$$\mathbf{Q}, \mathbf{Q}_p, \mathbf{R}_q, \mathbf{Z}_q, \mathbf{Q}_q \ (q \neq p).$$

Since $\dim L_p \leqslant 2$ and $\dim_{\mathbf{Z}_p} \leqslant \dim_{\mathbf{R}_p}$ (inequality (15)), in order to establish (26), it suffices to show that $\dim_{\mathbf{Z}_p}(L_p) \geqslant 2$. It is easy to verify that, upon taking the polyhedron L to be the sphere S^2 and attaching a single Moebius band M_p as described above, we obtain

$$H^2(M_p(S^2); \mathbf{Z}) = \mathbf{Z}_p; \qquad H^1(M_p(S^2); \mathbf{Z}) = 0. \qquad (28)$$

Note that $M_2(S^2) \cong \mathbb{R}P^2$.

Set $U^1 = M_p(\bar{T}^2) \backslash S^1 \cong M_p(S^2) \backslash \{a\}$ where $a \in M_p(S^2)$ is an arbitrary point. Writing out the exact sequence of the pair $(M_p(S^2), \{a\})$, and using (28), we obtain

$$H_c^2(U^1; \mathbf{Z}) = \mathbf{Z}_p, \ H_c^1(U^1; \mathbf{Z}) = 0. \qquad (29)$$

Now observe that for $k > 0$ the polyhedron $L_{k,p}$ is homeomorphic to a polyhedron $M_p(L)$ where $L = \bar{T}^2$ is a triangle taken with finer and finer triangulation as k grows larger. If $M_p(L)$ is obtained from $\bar{T}^2$ by attaching r copies of M_p, then we denote it by M_p^r. It is clear that $M_p^1 = M_p$. Set $U^r = M_p^r \backslash S^1$ where S^1 is the boundary of $\bar{T}^2$. We show by induction on r that

$$H_c^2(U^r, \mathbf{Z}) = \mathbf{Z}_p. \qquad (30)$$

For $r = 1$, equality (30) has already been verified. We can represent U^r as a union of U^{r-1} and U^1 where the intersection $U^{r-1} \cap U^1$ is homeomorphic to an open disk U (see Fig. 11 where the black dots correspond to disks which have been replaced by Moebius bands).

From the exactness with respect to the addition sequence of a triad

$$H_c^2(U^{r-1} \cap U^1) \to H_c^2(U^{r-1}) \oplus H_c^2(U^1) \to H_c^2(U^r) \to H_c^3(U^{r-1} \cap U^1)$$

$$\| \qquad\qquad \| \qquad\qquad \|$$

$$\mathbf{Z} \qquad\qquad \mathbf{Z}_p \oplus \mathbf{Z}_p \qquad\qquad 0$$

we get three possibilities for $H_c^2(U^r, \mathbf{Z})$: 0, $\mathbf{Z}_p$, and $\mathbf{Z}_p \oplus \mathbf{Z}_p$. But $H_c^2(U^r) \neq 0$ since $\mathbf{Z}$ cannot be mapped epimorphically onto $\mathbf{Z}_p \oplus \mathbf{Z}_p$. On the other hand, all two dimensional simplices of the triangulated polyhedron M_p^2 are cohomologous to one another. Thus, $H_c^2(U^r) \cong H^2(M_p^r(S^2))$ is a quotient group of $\mathbf{Z}$ and (30) follows.

From (30), on the one hand, we easily obtain (26). On the other hand, by the universal coefficient formula,

$$H_c^2(U^r; G) = 0 \qquad (31)$$

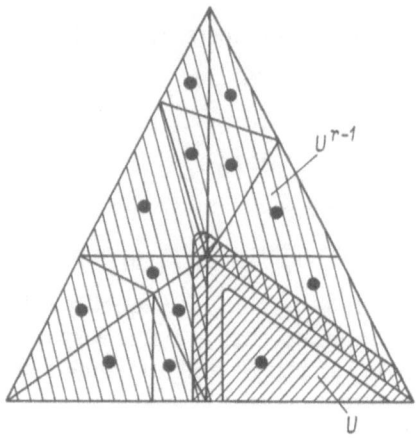

Fig. 11

for any of the following groups: $\mathbb{Q}$, $\mathbb{Q}_p$, $\mathbb{R}_q$, $\mathbf{Z}_q$, $\mathbb{Q}_q$ $(q \neq p)$. This, in turn, easily implies (27).

Then for $p \neq q$ we have $\dim(L_p \times L_q) = \dim_{\mathbf{Z}}(L_p \times L_q) = $ (by (19)) = $\max\{\dim_{\mathbb{R}_r}(L_p \times L_q);\ r\ \text{prime}\}$. But, it follows from Theorem 10 that $\dim_{\mathbb{R}_p}(L_p \times L_q) = \dim_{\mathbb{R}_q}(L_p \times L_q) = 3$ and $\dim_{\mathbb{R}_r}(L_p \times L_q) = 2$ for $r \neq p,\ q$. Thus, $\dim(L_p \times L_q) = 3$.

4.3. Dimensionally Full-Bodied Compacta

Definition 3. A space X is said to be dimensionally full-bodied if $\dim(X \times Y) = \dim X + \dim Y$ for any compactum Y.

It follows from Theorem 10 and Corollary 3 that any one dimensional compactum is dimensionally full-bodied. Pontryagin's compacta were the first examples of compacta which were not dimensionally full-bodied. Boltyanskiĭ modified Pontryagin's construction to obtain two-dimensional compacta B_p for each prime number p for which $\dim(B_p \times B_p) = 3$. These compacta have the following cohomological dimensions

$\mathbb{R}_p$	$\mathbf{Z}_p$	$\mathbb{Q}_p$	$\mathbb{Q}$	$\mathbb{R}_q$	$\mathbf{Z}_q$	$\mathbb{Q}_q$
2	1	1	1	1	1	1

Theorem 11 ([Bo]). *A finite dimensional compactum X is dimensionally full-bodied if and only if* $\dim_G X = \dim X$ *for any group G.*

The sufficiency of the conditions follows from Theorem 10 and Corollary 3 in an obvious manner. We shall prove necessity. Suppose that $\dim(X \times B_p) = \dim X + 2$ for all p. Since $\dim_{\mathbb{R}_q}(X \times B_p) < \dim X + 2$ for $p \neq q$, we have $\dim_{\mathbb{R}_p}(X \times B_p) = \dim X + 2$. By Theorem 10, this is only possible if $\dim_{\mathbb{Q}_p} X =$

$\dim_{\mathbb{R}_p} X = \dim X$. It follows from Bokshteĭn's inequalities that $\dim_G X = \dim X$ for groups in the collection σ and, hence, by Theorem 9, for any group G.

It follows from the proof of Theorem 11 that a sufficient (and obviously necessary) condition for a compactum to be dimensionally full-bodied is that $\dim(X \times B_p) = \dim X + 2$ for all p. The following question of Bokshteĭn's is still unsolved.

Question 1. Does there exist a neighbourhood retract which is not dimensionally full-bodied?

§ 5. The Cohomological Dimension of Paracompact Spaces

In this section we shall only survey the main results of the theory of cohomological dimension of paracompact spaces. The proofs, which the reader can find in [K2], usually use techniques from the theory of sheaves, which clearly lie outside the framework of this essay.

5.1. Definitions and Simplest Properties of Cohomological Dimension

Definition 4. The *cohomological dimension* $\dim_G X$ of a paracompact space X with respect to a coefficient group G is the largest integer n for which there exists a closed subset $A \subset X$ such that the spectral cohomology group $H^n(X, A; G)$ based on the collection of all open covers of X is different from zero.

In [A2], Aleksandrov defined the cohomological dimension $\text{A-}\dim_G X$ of a normal space X using spectral cohomology based on finite open covers.

It is easy to establish that

$$\text{A-}\dim_G X = \text{A-}\dim_G \beta X = \dim_G \beta X \tag{32}$$

Significantly deeper is the fact that

$$\text{A-}\dim_G X = \dim_G X \tag{33}$$

when X is a finite-dimensional paracompact space X and G a finitely generated group. The equalities (32), (33) and (11) imply that $\dim_{\mathbb{Z}} X = \dim X$ for finite dimensional paracompact spaces X, a result obtained by Dowker in [D].

Definition 5. The *relative cohomological dimension* of a subset $A \subset X$ with respect to a coefficient group G is the number $\text{r-}\dim_G A = \sup\{\dim_G B : B$ is contained in A and closed in $X\}$.

Theorem 12. *If A is a closed subset of a paracompact space X or any subset of a hereditarily paracompact space X, then*

$$\dim_G A \leqslant \dim_G X.$$

Corollary 4. *For any subset A of a paracompact space X,*

$$\text{r-}\dim_G A \leqslant \dim_G X.$$

5.2. The Sum and Obstruction Theorems

Theorem 13. *Let* $\{A_\alpha\}$ *be a cover of a paracompact space* X. *The inequality*

$$\dim_G X \leqslant \max\{r\text{-}\dim_G A_\alpha\}$$

holds in each of the following cases:
1) *the* A_α *are open sets*;
2) *the* A_α *are closed and the cover* $\{A_\alpha\}$ *is locally countable*;
3) *the* A_α *are locally closed and the cover* $\{A_\alpha\}$ *is locally finite*.

Theorems 8 and 13 imply that Definitions 1 and 4 of the cohomological dimension coincide for locally compact paracompact spaces.

Definition 6. A *wealthy* base of a paracompact space X is a collection $\{U_\alpha\}$ of open subsets of X such that, for each closed subset A of X and each neighbourhood OA, there exists a locally finite (in X) cover of A by elements of $\{U_\alpha\}$ contained in OA.

Theorem 14. *Let* $\{U_\alpha\}$ *be a wealthy base of a paracompact space* X. *If* $\dim_G X = r$, *the group* $H^r(X, X\setminus U_\alpha; G)$ *is different from zero for some* U_α.

Using this theorem and duality between Steenrod-Sitnikov homology and spectral cohomology, it is easy to deduce the following result due to Sitnikov [S].

Theorem 15. *If* $X \subset \mathbb{R}^n$ *is a set of dimension* $\dim_G X = r$ *and* $V \subset \mathbb{R}^n$ *is any ball, then* $H_k(V\setminus X; G) = 0$ *for* $k < n - r - 1$; *moreover, for any* $\varepsilon > 0$, *there exists a ball* V_0 *of radius less than or equal to* ε *such that* $H_{n-r-1}(V_0\setminus X; G) \neq 0$.

5.3. Relations Between Dimensions with Respect to Different Coefficient Groups. The Dimension of a Product.
Bockstein's inequalities (12)–(16) hold for any paracompact space. In fact, they can be sharpened. Let $\mathbb{R}_p^*$ denote the ring of integer p-adic numbers. Since $\mathbb{R}_p^*$ is a module over the ring $\mathbb{R}_p$ we have

$$\dim_{\mathbb{R}_p^*} X \leqslant \dim_{\mathbb{R}_p} X. \tag{34}$$

For locally compact spaces, the opposite inequality also holds. The question of whether the dimensions with respect to $\mathbb{R}_p$ and $\mathbb{R}_p^*$ coincide in general remains open. In view of (34), the following inequalities sharpen the corresponding Bokshteĭn inequalities:

$$\dim_\mathbb{Q} X \leqslant \dim_{\mathbb{R}_p^*} X, \tag{14'}$$

$$\dim_{\mathbb{Z}_p} X \leqslant \dim_{\mathbb{R}_p^*} X, \tag{15'}$$

$$\dim_{\mathbb{Q}_p} X \leqslant \max\{\dim_\mathbb{Q} X, \dim_{\mathbb{R}_p^*} X - 1\}. \tag{16'}$$

The analogue of Theorem 10 is the following.

Theorem 16. *If* X *is compact and* Y *paracompact, then equalities* (21')–(25'), *obtained by replacing the ring* $\mathbb{R}_p$ *by* $\mathbb{R}_p^*$ *in* (21)–(25), *hold.*

In contrast to §4, we cannot directly apply this theorem to compute the dimension $\dim(X \times Y)$ because we do not know whether it is true that $\dim_Z X = \max_p \dim_{\mathbb{R}_p} X$ and $\dim_Z X = \max_p \dim_{\mathbb{R}_p^*} X$. As a result, the following turns out to be more suitable for applications.

Theorem 17. *Suppose that X and Y are finite dimensional spaces which are compact and paracompact, respectively. Then*

$$\max_p \dim_{\mathbb{R}_p^*} (X \times Y) \leqslant \dim(X \times Y) \leqslant \max_p \dim_{\mathbb{R}_p^*} (X \times Y) + 1$$

with equality

$$\dim(X \times Y) = \max \dim_{\mathbb{R}_p^*} (X \times Y) + 1$$

if and only if
  1) $\dim_{\mathbb{Q}}(X \times Y) = \dim_{\mathbb{R}_p^*} (X \times Y)$ *for every prime p,*
  2) $\dim_{\mathbb{Q}}(X \times Y) = \max \dim_{\mathbb{Q}_p}(X \times Y)$.

Corollary 5. *Let X be compact and Y paracompact and set $k = \min\{\dim_{\mathbb{Q}} Y, \min_p \dim_{\mathbb{Q}_p} Y\}$. Then $\dim(X \times Y) \geqslant \dim X + k$. In particular, if $\dim Y \geqslant 1$, then $\dim(X \times Y) \geqslant \dim X + 1$.*

However, it is not known whether $\dim(X \times Y) \geqslant \dim Y + 1$ if X is compact with positive dimension.

Theorem 18. *A finite dimensional paracompact space Y is dimensionally fullbodied if and only if $\dim Y = \dim_{\mathbb{Q}_p} Y$ for any prime p.*

Chapter 6
Dimension and Mappings

§1. The Characterization of Dimension in Terms of Maps

In Chapter 2, we presented characterizations of dimension by ε- and ω-maps, essential maps, and by extensions of maps to the sphere. In this section, we set forth some other characterizations.

Definition 1. A map $f: X \to Y$ is said to have *dimension less than or equal to k* (written $\dim f \leqslant k$) if $\dim f^{-1}y \leqslant k$ holds for every $y \in Y$.

Theorem 1 (Hurewicz). *Every n-dimensional metric compactum X can be mapped to the n-dimensional cube I^n by a zero-dimensional map.*

Moreover, the set of zero-dimensional maps $f: X \to I^n$ is an everywhere dense G_δ in the space $C(X, I^n)$.

Proposition 1. *If* $f: X \to Y$ *is a closed zero-dimensional map, then* ind $X \leqslant$ ind Y.

The proof is by induction on ind Y. Suppose that $x \in X$, $f(x) = y$ and O_x is any neighbourhood of x. If dim $f^{-1}y = 0$, then ind $f^{-1}y = 0$. Therefore, there exist open subsets U_1 and U_2 of X with disjoint closures such that $f^{-1}y \subset U_1 \cup U_2$ and $x \in U_1 \subset O_x$. Since f is closed, there exists a neighbourhood O_y such that $f^{-1}O_y \subset U_1 \cup U_2$ and ind Bd $O_y \leqslant$ ind $Y - 1$. Then the boundary of the set $V = U_1 \cap f^{-1}O_y$ is such that Bd $V \subset f^{-1}$ Bd O_y and, consequently, ind Bd $V \leqslant$ ind f^{-1} Bd $O_y \leqslant$ (by the induction assumption) $\leqslant$ ind Bd $O_y \leqslant$ ind $Y - 1$ from which it follows that ind $X \leqslant$ ind Y.

Theorem 1, Proposition 1 and the equality ind = dim for spaces with a countable base give the following result.

Theorem 2. *A necessary and sufficient condition that a metrizable compactum X have dimension* dim $X = n$ *is that there exist a zero-dimensional map $f: X \to I^n$ and no zero-dimensional map into a cube of lower dimension.*

We introduce the following generalization of the concept of an ω-map.

Definition 2. A map $f: X \to Y$ is called ω-*discrete* where ω is an open cover of X, if every point $y \in fX$ has a neighbourhood O_y in Y whose preimage $f^{-1}O_y$ is a discrete (in X) collection of open sets refining the cover ω.

Theorem 3. *A normal space X has dimension* dim $X \leqslant n$ *if and only if, for each open, locally finite cover ω, there exists an ω-discrete map of X into an n-dimensional simplex $\bar{T}^n$.*

The proof is obvious. Necessity follows from Proposition 8 of Chapter 3 and Theorem 5 of Chapter 2 about canonical maps.

Theorem 4 (Morita-Nagami). *If a normal space Y is the image of a zero-dimensional normal space X under a continuous closed map $f: X \to Y$ of multiplicity less than or equal to $n + 1$, then* Ind $Y \leqslant n$.

The proof is by induction on n. If F_1 and F_2 are disjoint closed subsets of Y, there exist open-closed disjoint subsets U_1 and U_2 of X such that $f^{-1}F_i \subset U_i$ and $U_1 \cup U_2 = X$. The set $V = Y \setminus fU_2$ is a neighborhood of F_1 such that Bd $V \subset fU_1 \cap fU_2$. Therefore, on the set $U_1 \cap f^{-1}$ Bd V, the map f has multiplicity less than $n + 1$. By the induction hypothesis, Ind Bd $V < n$, from which we obtain Ind $Y \leqslant n$.

Definition 3. A continuous map $f: X \to Y$ of a metric space X to a space Y is said to be *totally zero-dimensional* if, for every $\varepsilon > 0$ and any point $y \in f(x)$, there exists a neighbourhood O_y in Y whose preimage $f^{-1}O_y$ is a discrete (in X) collection of open sets of diameter less than ε.

It follows from Theorem 16 of Chapter 3 that any totally zero-dimensional map is zero-dimensional. It is not difficult to show that a closed zero-dimensional map of a metric space is totally zero-dimensional. In general, zero-dimensionality does not imply total zero-dimensionality as the following example shows. Let X

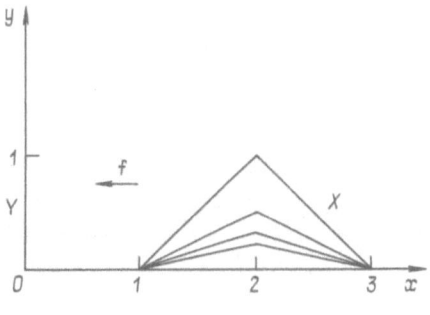

Fig. 12

be the subspace of the plane consisting of the countable set of segments joining the points $(1, 0)$ and $(3, 0)$ to the points $(2, 1/n)$, $n = 1, 2, \ldots$. Let f be the map consisting of horizontal projection to the segment $[(0, 0), (0, 1)] = Y$ of axis of the ordinate (see Fig. 12). It is not totally zero-dimensional at the point $(0, 0)$.

On the one hand, the following theorem characterizes the dimension in the spirit of Theorem 2 by means of zero-dimensional maps to a "better" space. On the other hand, it reveals a completely different aspect of n-dimensional spaces as finitely covered images of zero-dimensional maps.

Theorem 5. *For a metric space X the following conditions are equivalent:*
a) $\dim X \leqslant n$;
b) *the space X is the image of a metrizable space X^0 of dimension $\dim X^0 \leqslant 0$ and weight $wX^0 \leqslant wX$ under a closed map of multiplicity less than or equal to $n + 1$;*
c) *the space X admits a totally zero-dimensional map to a space Y with a countable base and dimension $\dim Y \leqslant n$.*

The implication b) $\Rightarrow$ a) follows from Theorem 4. To verify the implication c) $\Rightarrow$ a) it is sufficient, by Theorem 18c) of Chap. 3, to show that X is a union of $n + 1$ zero-dimensional sets. Let $Y = Y_0 \cup \cdots \cup Y_n$ where $\dim Y_i \leqslant 0$ and set $X_i = f^{-1}Y_i$ where $f: X \to Y$ is a totally zero-dimensional map. It is easily verified that, for every $\varepsilon > 0$, X_i admits a disjoint ε-cover. Hence $\dim X_i \leqslant 0$ by Theorem 16 of Chap. 3. To verify the implication a) $\Rightarrow$ c) we first show that there exists a totally zero-dimensional map $f: X \to Q$. Let ω_i be the cover consisting of balls of radius $(\frac{1}{2})^i$, and let $f_i: X \to \bar{T}^n$ be the ω_i-discrete map which exists by Theorem 3. Take f to be the diagonal product of the f_i. Then apply the factorization Theorem 23 of Chap. 3 using the fact that when a totally zero-dimensional map f is a composition $f = h \circ g$, the map g is also totally zero-dimensional.

We first verify the implication a) $\Rightarrow$ b) under the assumption that, for each n-dimensional space Y with a countable base, there exists a closed epimorphism $\varphi: Y^0 \to Y$ of multiplicity less than or equal to $n + 1$ of a zero-dimensional space Y^0 with a countable base onto Y. By the (already established) implication a) $\Rightarrow$ c), there exists a zero-dimensional map $f: X \to Y$, where Y is an n-dimensional space with a countable base. Putting X^0 equal to the fan product of X and Y^0 with respect to the maps f and φ (see §7, article I), we obtain the desired map $\pi: X^0 \to X$ parallel to φ. Concerning spaces with a countable base, Theorem 19 of Chapter 2 implies that it is sufficient to verify the implication a) ⇒ c) for compact X. Take a refining sequence of closed finite covers $\alpha_m = \{F_1^m, \ldots, F_{k(m)}^m\}$ of the compactum X of multiplicity less than or equal to $n + 1$ so that α_{m+1} is a refinement of α_m. Put $X_m = \bigcup \{F_j^m \times \{j\}: 1 \leqslant j \leqslant k(m)\}$. There exist natural maps $f_m^{m+1}: X_{m+1} \to X_m$. Set $X_m^0 = \varprojlim \{X_m: f_m^{m+1}, m = 1, 2, \ldots\}$. Then the limiting projection $f: X^0 \to X$ is the desired $(n + 1)$-fold map of a zero-dimensional space X^0 to X.

§2. Closed Maps which Raise Dimension

Theorem 6. *If a closed map $f: X \to Y$ of a normal space X to a normal space Y has multiplicity less than or equal to $k + 1$, then*

$$\dim Y \leqslant \dim X + k. \tag{1}$$

This theorem was proved by Hurewicz for spaces with a countable base, by Morita for metrizable spaces, and by Zarelya in general. The inequality

$$\dim Y \leqslant \operatorname{Ind} X + k$$

is proved by induction on $\operatorname{Ind} X + k = n$. Passage from $n - 1$ to n is verified by induction on $\operatorname{Ind} X = r$. The case $r = 0$ was considered in Theorem 4. Let F_1 and F_2 be disjoint closed subsets of Y and choose a subset C of X separating $f^{-1}F_1$ and $f^{-1}F_2$ of dimension $\operatorname{Ind} C \leqslant r - 1$. There exist closed subsets $\Phi_i \supset F^{-1}F_i$ of X such that $C = \Phi_1 \cap \Phi_2$ and $X = \Phi_1 \cup \Phi_2$. It is easy to prove that if $D = f\Phi_1 \cap f\Phi_2$, then

$$\operatorname{rd}_D(D \setminus fC) \leqslant n - 1.$$

By the induction assumption, $\dim fC \leqslant n - 1$. Consequently, we have $\dim D \leqslant n - 1$ by Theorem 14 of Chap. 2. Thus, D separates F_1 and F_2 and has dimension $\dim D \leqslant n - 1$. Hence, $\dim Y \leqslant n$.

Replacing the reference to Theorem 14 of Chapter 2 by a reference to Theorem 28 of Chap. 3 in the above argument proves the following.

Theorem 7 (Morita). *If a continuous closed map $f: X \to Y$ of a normal space X to a perfectly normal space Y has multiplicity less than or equal to $k + 1$, then*

$$\operatorname{Ind} Y \leqslant \operatorname{Ind} X + k. \tag{2}$$

Remark 1. For maps of finite dimensional compacta the analogue of formula (1) is given by replacing the dimension dim by the cohomological dimension $\dim_G$ with respect to any coefficient group.

§3. Maps which Lower Dimension

Hurewicz obtained the formula

$$\dim X \leqslant \dim f + \dim Y \qquad (3)$$

for closed maps of spaces with a countable base. Morita generalized this to

$$\dim X \leqslant \dim f + \operatorname{Ind} Y \qquad (4)$$

for closed maps of a normal space X to a paracompact space Y. Sklyarenko proved inequality (3) for paracompact spaces X and Y. Later Pasynkov proved that paracompactness of X can be weakened to normality. Finally, Pasynkov and Filippov proved that (3) holds for a closed map of a normal space X to a weakly paracompact space Y.

It is absolutely impossible to waive the requirement that Y satisfy some property akin to paracompactness. Filippov constructed an example of a closed zero-dimensional map $f: X \to Y$ of a one-dimensional space X to a zero-dimensional space Y. Fedorchuk strengthened Filippov's example of a violation of (3). He proved, assuming Jensen's combinatorial principle, that a zero-dimensional perfect map to a perfectly normal, countably compact space can lower dimension. He constructed maps which simultaneously lower the dimensions dim and Ind and maps which only lower dim. Constructing such examples without additional set-theoretic hypotheses is impossible in view of Weiss's theorem about the compactness of perfectly normal, countably compact spaces if one assumes Martin's axiom and the negation of the continuum hypothesis.

We now prove (3) for metric spaces X and Y. We first consider the case where Y is zero-dimensional and show that $\dim X = \dim f = n$. Let ω be an open cover of X. For each $y \in Y$ there exists a collection u_y of open subsets of X which covers $f^{-1}y$, refines ω and has multiplicity less than or equal to $n + 1$. Since f is closed, there exists a neighbourhood O_y of y such that $f^{-1}O_y \subset \tilde{u}_y$. Refine $\{O_y : y \in Y\}$ by a disjoint open cover v. For every $V \in v$, choose a point $y \in Y$ such that $f^{-1}V \subset \tilde{u}_y$ and put $u_v = \{U \cap f^{-1}V : U \in u_y\}$. Then $u = \bigcup\{u_v : V \in v\}$ is an open cover of X which refines the cover ω and has multiplicity less than or equal to $n + 1$.

Now let $\dim Y = m > 0$. According to Theorem 5 there exists a closed $(m + 1)$-fold map $g: Y^0 \to Y$ of a zero-dimensional metric space Y^0 to Y. Consider the commutative diagram

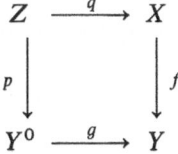

in which Z is the fan product of X and Y^0 with respect to the maps f and g and p and q are the projections "parallel to" f and g. By the "parallels lemma" (see §7, Article I), the maps p and q are closed, $\dim p = \dim f$ and the multiplicity of

q coincides with the multiplicity of g. By the zero-dimensional case, we have $\dim Z = \dim f = n$ and, consequently, $\dim X \leqslant m + n$ by Theorem 6.

Definition 4. We say that a *map* $f: X \to Y$ is *parallel to a space* Z (written $f \| Z$) if there exists an imbedding $X \subset Y \times Z$ such that $f = p_1 | X$ where $p_1: Y \times Z \to Y$ is projection. In this case, we shall also say that the map f is *imbedded in a projection parallel to* Z.

A very strong result characterizing k-dimensional perfect mappings of metric spaces was obtained by Pasynkov.

Theorem 8. *A perfect map* $f: X \to Y$ *of finite-dimensional metric spaces is k-dimensional if and only if* $f = f_k \ldots f_1 f_0$ *where* $\dim f_0 = 0$ *and each* $f_i, i = 1, \ldots, k$, *is parallel to the segment for* $i = 1, \ldots, k$.

From Theorem 8 we obtain the following.

Corollary 1. *A perfect map* $f: X \to Y$ *of finite-dimensional metric spaces is k-dimensional if and only if it is a composition of k one-dimensional maps.*

The analogue of formula (3) also holds for cohomological dimensions.

Theorem 9. *If* $f: X \to Y$ *is a map of a finite-dimensional compactum X to a finite-dimensional compactum Y, then*

$$\dim_G X \leqslant \dim f + \dim_G Y, \tag{5}$$

$$\dim_G X \leqslant \dim_G f + \dim Y, \tag{6}$$

$$\dim_G X \leqslant \dim_G f + \dim_G Y + 1, \tag{7}$$

for any coefficient group. If the group G is such that, for each prime number p, $Q_p \in \sigma(G)$ implies either $\mathbf{Z}_p \in \sigma(G)$ *or* $\mathbb{R}_p \in \sigma(G)$ *(see Chap. 5), then*

$$\dim_G X \leqslant \dim_G f + \dim_G Y. \tag{8}$$

Definition 5. A continuous map $f: X \to Y$ is called *fully closed* if, for each point $y \in Y$ and each finite collection $\{U_1, \ldots, U_s\}$ of open subsets of X covering $f^{-1}y$, the set $\{y\} \cup (\bigcup \{f \# U_i: i = 1, \ldots, s\})$ is open.

If we demand that $s = 1$ in this definition, then we obtain an equivalent definition of a closed map. Although the class of fully closed maps seems rather specialized (for example, the projection of the square onto the segment is not a fully closed map), it plays an important role in the theory of dimension. This stems from the fact that fully closed maps do not decrease dimension. More precisely, the following strengthened version of Hurewicz's formula

$$\dim X \leqslant \max\{\dim f, \dim Y\} \tag{9}$$

holds for fully closed maps f. For fully closed mappings to a paracompact space, this result was obtained by Fedorchuk. Savinov proved that paracompactness of Y can be dropped in this case. He also proved that $\dim Y \leqslant \dim X + 1$ if

$f: X \to Y$ is a fully closed map. If, in addition, X is perfectly normal, then

$$\text{Ind } Y \leqslant \text{Ind } X + 1, \tag{10}$$

$$\text{Ind } X \leqslant \text{Ind } f + \text{Ind } Y. \tag{11}$$

§4. Open Maps which Raise Dimension

4.1. Open Maps with Countable Multiplicity. In this section, we shall consider open maps which do not raise the dimension. A map $f: X \to Y$ is said to be of *countable multiplicity* if the pre-image of $f^{-1}y$ of any point $y \in Y$ is at most countable. A map $f: X \to Y$ is said to be *local homeomorphism at a point $x \in X$* if there exists a neighbourhood O_x of x on which the restriction of f is a homeomorphism. It is clear that the set of points at which f is a local homeomorphism is open in X.

The following theorem was proved by Kolmogorov for metrizable compacta and by Pasynkov in general.

Theorem 10. *If X is a locally compact space, the set of points at which a map $f: X \to Y$ of countable multiplicity is a local homeomorphism is everywhere dense in X.*

Thus, for example, the reader can easily verify that the set of points at which an open map $f: I \to I$ is not a local homeomorphism is finite and that the map itself has the form

$$f(x) = \tfrac{1}{2}[\sin(k\pi/2 + n\pi\varphi(x)) + 1]$$

where $\varphi: I \to I$ is an orientation preserving homeomorphism, $k = \pm 1$, and n is a natural number.

It follows from Theorem 5 that every finite dimensional metrizable compactum is the image of a zero-dimensional space under a map of finite multiplicity; that is, a (closed) map of finite multiplicity can raise dimension. This is not the case for open mappings.

Theorem 11. *If an open map $f: X \to Y$ of a compactum X to a compactum Y has countable multiplicity, then $\dim Y = \dim X$. If, in addition, Y is perfectly normal, then $\text{Ind } X = \text{Ind } Y$.*

This theorem was obtained by Aleksandrov in the metrizable case, and by Pasynkov in general. The first assertion of the theorem is proved as follows. By transfinite recursion based on Theorem 10, one constructs a sequence of open subsets O_α of Y such that

a) loc dim $O_\alpha \leqslant \dim X$;

b) $O_\alpha \subset O_{\alpha+1}$;

c) if $y \backslash O_\alpha \neq \varnothing$, then $O_{\alpha+1} \backslash O_\alpha \neq \varnothing$.

Suppose that we have already constructed O_α. The map $f_\alpha = f|_{X \backslash f^{-1}O_\alpha}$ is open. According to Theorem 10, the set $U_\alpha \subset X \backslash f^{-1}O_\alpha$ of points at which f_α is a local homeomorphism is not empty. Set $O_{\alpha+1} = O_\alpha \cup fU_\alpha$. Then $Y = O_\alpha$ for some α_0 and, by Theorem 13 of Chapter 3, we have

$$\dim Y = \text{locdim } O_{\alpha_0} \leqslant \dim X.$$

The inequality $\dim X \leqslant \dim Y$ follows from formula (3). The second assertion of the theorem can be proved in a similar way because, by Theorem 28 of Chapter 3, Ind $Y = \text{loc Ind } Y$ when Y is a perfectly normal compactum.

4.2. The Examples of Kolmogorov and Keldysh. In Theorem 11, it is not possible to replace countable multiplicity of the map f by zero-dimensionality. Kolmogorov constructed the first example of zero-dimensional open map of metrizable compactum X to a compactum of larger dimension. His example maps a one-dimensional compactum to Pontryagin's example of a two-dimensional compactum L_p which is not dimensionally full-bodied (see Chap. 5, §4).

An example of a zero-dimensional, open map of a one-dimensional compactum to the square was constructed by Keldysh. Keldysh's construction was very complicated. We present a significantly simpler version due to Kozlovskiĭ.

Let S_1^1 and S_3^1 be concentric circles of radius 1 and 3, respectively, on the plane. Parametrize the points of S_1^1 and S_3^1 by the angle φ, $0 \leqslant \varphi < 2\pi$. Let $_\varphi S_2^1$ be the circle of radius 2 inside S_3^1 which is tangent to S_3^1 at the point φ and S_1^1 at the point $\varphi + \pi$ (mod 2π). Inside the disk bounded by S_1^1, draw a horizontal chord H_φ through the point of tangency of $_\varphi S_2^1$ and S_1^1 (see Fig. 13). We put $K_\varphi = \varphi S_2^1 \cup H$ (this set is outlined more heavily in Fig. 13).

We now construct a triple-valued retraction r of the disk B_3^2 bounded by S_3^1 to S_3^1. To each point $x \in B_3^2$ associate the points $\varphi \in S_3^1$ for which $x \in K_\varphi$. It is easy to verify that this map is continuous and at most triple-valued. The projection $g_1 \colon \Phi_1 \to B_3^2$ of its graph Φ_1 is an open, triple-valued, inessential map to the two dimensional disk.

Thinking of the parameterized family of circles $_\varphi S_2^1$ as a torus, the chords H_φ rotate inside this torus making a complete turn and twice degenerating into a

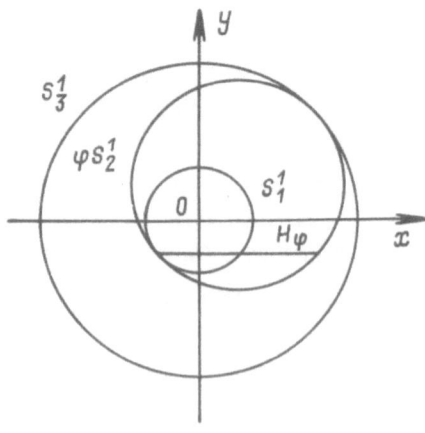

Fig. 13

point. The compactum Φ_1 is a curvilinear polyhedron which can be thought of as this torus omitting the interior of the twisted tape formed by the chords H_φ. The map g_1 consists of "laying" the meridianal compacta K_φ of the polyhedron Φ_1 onto the plane of the disk B_3^2.

Similarly, take any triangulation of the polyhedron Φ_1 and replace every two-dimensional simplex of the triangulation by a copy of the polyhedron Φ_1 to obtain a two-dimensional polyhedron Φ_2 mapping onto Φ_1 by means of an open map g_2 of multiplicity three sweeping out onto the one-dimensional skeleton of the polyhedron Φ_1. Continuing this process, we obtain an inverse system $S = \{\Phi_i, g_i\}$ of two-dimensional polyhedra and open adjacent projections $g_i \colon \Phi_i \to \Phi_{i-1}$ of multiplicity three sweeping out onto the one dimensional frame of the polyhedron Φ_{i-1}. The limit of this spectrum under successive refinement of the triangulations of the polyhedra Φ_i is a one-dimensional compactum which is mapped by an open, zero-dimensional mapping onto the two-dimensional disk.

4.3. Compacta as Open Images of One-Dimensional Compacta. Using Keldysh's example, Pasynkov obtained the following general result.

Theorem 12. *Every non-zero-dimensional compactum Y of weight τ is the image of a one-dimensional compactum $X = X(Y)$ of weight τ under an open zero-dimensional map $f \colon X \to Y$.*

This result cannot be strengthened in the class of compacta, since it is obvious that the open image of a zero-dimensional compactum is zero-dimensional. The proof of Theorem 12 is based on the following two lemmas.

Lemma 1. *Every two-dimensional metrizable compactum Y is an open zero-dimensional image of a one-dimensional metrizable compactum.*

Lemma 2. *If $f_n \colon X_n \to I^n$ is a zero-dimensional map of a one-dimensional metrizable compactum X_n to the n-dimensional cube I^n, then there exists a one-dimensional compactum X_{n+1} and open maps $f_{n+1} \colon X_{n+1} \to I^{n+1}$, $q_n^{n+1} \colon X_{n+1} \to X_n$ such that f_{n+1} is zero-dimensional and the diagram*

$$
\begin{array}{ccc}
X_{n+1} & \xrightarrow{\ f_{n+1}\ } & I^{n+1} \\[2pt]
{\scriptstyle q_n^{n+1}}\Big\downarrow & & \Big\downarrow{\scriptstyle p_n^{n+1}} \\[2pt]
X_n & \xrightarrow{\ f_n\ } & I^n
\end{array}
\tag{12n}
$$

is commutative where p_n^{n+1} is the natural projection and, moreover, for any point $x \in X_n$, we have

$$
f_{n+1}(q_n^{n+1})^{-1}(x) = (p_n^{n+1})^{-1} f_n x. \tag{13}
$$

The map $f\colon X \to Y$ for Lemma 1 is taken from the commutative diagram

$$
\begin{array}{ccc}
X & \xrightarrow{\ f\ } & Y \\
\downarrow{\scriptstyle p} & & \downarrow{\scriptstyle h} \\
K & \xrightarrow{\ h\ } & I^2
\end{array}
$$

and is the fan product of the open zero-dimensional map g constructed by Keldysh of the one dimensional compactum K to the square and an essential, by Theorem 1, zero-dimensional map $h\colon Y \to I^2$. The maps p and f are zero-dimensional and f is open by the "parallels" lemma (see §7, Article I). The compactum X is one dimensional by formula (3).

To prove Lemma 2, we consider the commutative diagram

$$
\begin{array}{ccc}
X & \xrightarrow{\ g\ } & I^{n+1} \\
\downarrow{\scriptstyle h} & & \downarrow{\scriptstyle p_n^{n+1}} \\
X_n & \xrightarrow{\ f_n\ } & I^n
\end{array}
$$

which is the fan product of the maps f_n and p_n^{n+1}. The compactum Y has dimension $\dim Y \leqslant 2$ by formula (3). Then we set $X_{n+1} = X$, $f_{n+1} = g \circ f$, $q_n^{n+1} = h \circ f$, where $f\colon X \to Y$ is the map from Lemma 1.

The diagrams (12n) form two parallel inverse systems $S = \{X_n, q_n^{n+1}\}$ and $T = \{I^n, p_n^{n+1}\}$ and a morphism $F = \{f_n\}\colon S \to T$ of one to the other. The map $f = \lim F\colon \varprojlim S \to \varprojlim T = Q$ is an open (by condition (13)) zero-dimensional map of a one dimensional compactum to the Hilbert cube. Theorem 12 is proved by transfinite induction on τ using the corresponding analogs of Lemmas 1 and 2.

Using a similar approach, Fedorchuk proved the following result.

Theorem 13. *Every compactum Y of weight τ is the image of a one-dimensional compactum $X = X(Y)$ of weight τ under an open monotone map $f\colon X \to Y$.*

Instead of using Keldysh's example as a starting point, Theorem 13 uses an example due to Anderson of monotone open map $f\colon M_1^3 \to Q$ of the universal Menger curve to the Hilbert cube. Theorem 13 plays an essential role in the theory of extensors.

§5. Maps of Infinite Dimensional Spaces

5.1. A Characterization of Countable Dimensionality

Theorem 14. *A metric space X is countable dimensional if and only if it is the image of a zero-dimensional metric space X^0 under a closed map which is of*

a) *finite multiplicity,*
b) *countable multiplicity, or*
c) *multiplicity less than c.*

Nagata showed that condition a) follows from the countable dimensionality of X. The implications a) $\Rightarrow$ b) $\Rightarrow$ c) are trivial. That condition c) implies that X is countable dimensional was shown by Sklyarenko if X is a compactum, by Smirnov if X is complete, and by Arkhangel'skiĭ and Nagami in general.

5.2. Maps which Raise Dimension. The main questions here are to identify the types of maps which do not leave the class of countable dimensional spaces or spaces for which the dimensions ind and Ind are defined.

It easily follows from condition c) of Theorem 14 that the result below holds.

Theorem 15. *If $f: X \to Y$ is map of a countable dimensional metric space to an uncountable dimensional metric space Y, then the set Y_1 of those points $y \in Y$ for which the preimage $f^{-1}y$ has cardinality greater than or equal to c is not countable dimensional.*

Corollary 2. *The image of a countable dimensional metric space under a closed map of countable multiplicity is at most countable dimensional.*

Theorem 16. *Let $f: X \to Y$ be a closed map of bounded finite multiplicity from a normal space X to a hereditarily normal space Y. If the dimension Ind X (respectively, ind X) is well defined, then the dimension Ind Y (respectively, ind Y) is defined.*

The proof proceeds along the same lines as the proof of Theorem 7, except that the following proposition is used instead of Theorem 14 of Chap. 2.

Proposition 2. *Let X be a hereditarily normal space and A a closed subset such that the transfinite dimensions Ind A and r Ind$(X\backslash A)$ (respectively, ind A and r ind$(X\backslash A)$) are defined. Then the transfinite dimension Ind X (respectively, ind X) is defined.*

Remark 1. In Theorem 16, it is not possible replace the condition that f have bounded finite multiplicity by the condition that it have finite multiplicity. Indeed, Theorem 5 implies that there exists an $(n + 1)$-dimensional map $f_n: X_n \to I^n$ of a zero-dimensional compactum X_n to the n-dimensional cube for each $n > 0$. Putting $X = \bigoplus_n X_n$, $Y = \bigoplus_n I^n$ and $f = \bigoplus_n f_n$ gives a map $f: X \to Y$ of finite multiplicity from a zero-dimensional space X to a space Y whose dimension Ind Y is not defined (see Chap. 4).

Concerning the dimension ind, by Theorem 14 every countable dimensional (separable) metric space is the image of some zero-dimensional space under a closed map of finite multiplicity. This is true, in particular, for the universal countable dimensional separable metric space (see §1, Chap. 4) which, because it contains compacta of all countable transfinite dimensions (by Theorem 1, Chap. 4), cannot have transfinite dimension ind.

The analogue of Theorem 11 for transfinite dimensions is the following.

Theorem 17 (Fedorchuk). *If $f: X \to Y$ is an open map of countable multiplicity from a Čech complete (in particular, locally compact) space X to a hereditarily normal space Y and if the dimension Ind X is defined, then the dimension ind Y is defined.*

From this theorem and Theorem 2 of Chap. 4 we immediately obtain the following.

Corollary 3. *If $f: X \to Y$ is an open map of countable multiplicity from a locally compact space X to a hereditarily normal compactum Y and if the dimension ind X is defined, then the dimension Ind Y is defined.*

Theorem 18 (Fedorchuk). *If $f: X \to Y$ is an open map of countable multiplicity from a Čech complete space X to a hereditarily normal paracompact space Y and if the dimension Ind X is defined, then the dimension Ind Y is also defined.*

5.3. Maps Lowering Dimension. The main result of this section is the following.

Theorem 19 (Levshenko). *Let $f: X \to Y$ be a map of compacta and suppose that the dimension ind Y is defined and all fibers $f^{-1}y$ are weakly infinite dimensional. Then the compactum X is also weakly infinite dimensional.*

Question 1. Is it sufficient in the formulation of Theorem 19 to require that the compactum Y be weakly infinite dimensional?

It is worth remarking that a positive answer to Question 1 would give a positive solution to Question 7 of Chap. 4 about the product of weakly infinite dimensional compacta. A partial answer to Question 1 was given by V.V. Fedorchuk.

Theorem 20. *Let $f: X \to Y$ be a fully closed map with weakly infinite-dimensional fibers $f^{-1}y$ of a compactum X to a weakly infinite-dimensional compactum Y. Then the compactum X is also weakly infinite dimensional.*

In addition, it can be shown that the limit of an inverse sequence of weakly infinite dimensional compacta and completely closed projections is weakly infinite-dimensional. This result appeared in connection with the solution of the problem of constructing hereditarily uncountable dimensional, weakly infinite dimensional compacta.

Theorem 19 does not carry over to countable dimensional compacta even in the metrizable case.

Example 1 (R. Pol). *There exists a map $f: X \to C$ of an uncountable dimensional metrizable compactum X to a Cantor perfect set, all fibers $f^{-1}c$ of which are countable dimensional. Moreover, the dimensions ind $f^{-1}c$ of the fibers are bounded from above by a countable transfinite α.*

Nevertheless, the following holds.

Theorem 21. *Let $f: X \to Y$ be a closed map of a metric space X to a countable dimensional metric space Y whose fibers $f^{-1}y$ are finite dimensional. Then the space X is also countable dimensional.*

Proof. By Theorem 14, there exists a closed map $g\colon Y^0 \to Y$ with finite multiplicity from a zero-dimensional metric space Y^0 to Y. Consider the commutative diagram

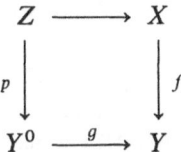

in which Z is a fan product of the spaces X and Y^0 with respect to the maps f and g and p and q are the natural projections onto the components. Then, by formula (3) in § 3, the space Z is countable dimensional and, by Corollary 2, the space X will also be countable dimensional.

Corollary 4. *Let $f\colon X \to Y$ be a map with finite dimensional fibers from a metrizable compactum X to a compactum Y and suppose the transfinite dimensions of Y are defined. Then the transfinite dimensions of the compactum X are also defined.*

Bibliography*

The bibliography below contains the original works on dimension theory by Poincaré [PO], Brouwer [B1, B2], and Lebesgue [L1, L2]. It also includes the monographs of Hurewicz and Wallman [HW], Nagami [N], Nagata [Na], Aleksandrov and Pasynkov [AP], Pears [P], and Engelking [E]. Among the other citations, it is worth mentioning Aleksandrov's article [A1] which presents the principles of the homological theory of dimension. For a fuller bibliography on dimension theory, the reader can consult the aforementioned monographs and the survey written by Pasynkov, Fedorchuk, and Fillipov [PFF].

[AP]   Aleksandrov, P.S., Pasynkov, B.A.: Introduction to Dimension Theory, Nauka, Moscow, 1973. 576pp. (Russian). Zbl. 272.54028
[A1]   Alexandroff, P. (= Aleksandrov, P.S., Alexandrov, P.S.): Dimensionstheorie. Ein Beitrag zur Geometrie der abgeschlossenen Mengen, Math. Ann. *106* (1932) 161–238. Zbl.4,73
[A2]   Alexandroff, P., On the dimension of normal spaces. Proc. R. Soc. Lond., Ser. A *189* (1947) 11–39. Zbl.38,361
[B0]   Boltyanskiĭ, V.G.: On the dimensional full valuedness of compacta, Dokl. Akad. Nauk. SSSR *67*, No. 5 (1949) 773–776 (Russian). Zbl.37,98
[B1]   Brouwer, L.E.J.: Beweis der Invarianz der Dimensionenzahl, Math. Ann. *70* (1911) 161–165
[B2]   Brouwer, L.E.J.: Über den natürlichen Dimensionsbegriff, J. Reine Angew. Math. *142* (1913) 146–152
[D]   Dowker, C.H.: Mapping theorems for noncompact spaces, Am. J. Math. *69* (1947) 200–242. Zbl.37,101
[E]   Engelking, R., Dimension Theory, Warsaw, 1978. 314pp. Zbl.401.54029

---

*For the convenience of the reader, references to reviews in Zentralblatt für Mathematik (Zbl.), compiled using the MATH database, and Jahrbuch über die Fortschritte der Mathematik (Jrb.) have, as far as possible, been included in this bibliography.

[F]   Fedorchuk, V.V.: Infinite Dimensional Bicompacta, Izv. Akad. Nauk SSSR, Ser Mat. *42*,
          No. 5 (1978) 1163–1178. Zbl.403.54025; English translation: Math. USSR, Izv. 13 (1979)
          445–460
[HW]   Hurewicz, W., Wallman, H.: Dimension Theory, Princeton, 1948. 165pp. Zbl.36,125
[K1]   Kuz'minov, V.I.: On the continua V^n, Dokl. Akad. Nauk SSSR *139*, No. 1 (1961) 24–27.
          Zbl.107,167; English translation: Sov. Math., Dokl. 2 (1961) 858–861
[K2]   Kuz'minov, V.I.: The homological theory of dimension, Usp. Mat. Nauk *23*, No. 5 (1968)
          3–49. Zbl.179,279; English translation: Russ. Math. Surv. 23, No. 5 (1968) 1–45
[L1]   Lebesgue, H.: Sur la non applicabilité de deux domains appartenant à des espaces de n et
          $n + p$ dimensions. Math. Ann. *70* (1911) 166–168
[L2]   Lebesgue, H.: Sur les correspondances entre les points de deux espaces. Fundam. Math. *2*
          (1921) 256–285. Jrb.48,652
[N]   Nagami, K.R.: Dimension Theory, Academic Press, 1970. XI, 256pp. Zbl.224.54060
[Na]   Nagata, J.: Modern Dimension Theory, North Holland, Amsterdam, 1965, 259pp. Zbl.129,
          383
[PFF]   Pasynkov, B.A., Fedorchuk, V.V., Filippov, V.V.: Dimension Theory, Itogi Nauki Tekh.,
          Ser. Algebra Topologiya Geom. *17* (1973) 229–306. Zbl.444.54024; English translation: J.
          Sov. Math. 18 (1982) 789–841
[P]   Pears, A.R.: Dimension Theory of General Spaces, Cambridge Univ. Press, Cambridge,
          1975. 428pp. Zbl.312.54001
[Po]   Poincaré, H.: Pourquoi l'espace à trois dimensions? Rev. Metaphys. Morale, *20* (1912) 486p.
[S]   Sitnikov, K.A.: Combinatorial topology of nonclosed sets. I, *Mat. Sb.*, Nov. Ser. *34*, No. 1
          (1954) 3–54 (Russian). Zbl.55,163
[Z]   Zarelya, A.V.: Methods of the theory of rings of functions for the construction of compact
          extensions (Russian), in the book: Contrib. Extens. Theory Topol. Struct., Proc. Sympo,
          Berlin 1967, Berlin, (1969) 249–256. Zbl.183,513

Author Index

Subject Index

Encyclopaedia of Mathematical Sciences
Editor-in-chief: R. V. Gamkrelidze

Dynamical Systems

Volume 1: **D. V. Anosov, V. I. Arnol'd** (Eds.)
Dynamical Systems I
Ordinary Differential Equations and Smooth Dynamical Systems
1988. IX, 233 pp. ISBN 3-540-17000-6

Volume 2: **Ya. G. Sinai** (Ed.)
Dynamical Systems II
Ergodic Theory with Applications to Dynamical Systems and Statistical Mechanics
1989. IX, 281 pp. 25 figs.
ISBN 3-540-17001-4

Volume 3: **V. I. Arnol'd** (Ed.)
Dynamical Systems III
1988. XIV, 291 pp. 81 figs.
ISBN 3-540-17002-2

Volume 4: **V. I. Arnol'd, S. P. Novikov** (Eds.)
Dynamical Systems IV
Symplectic Geometry and its Applications
1989. VII, 283 pp. 62 figs.
ISBN 3-540-17003-0

Volume 5: **V. I. Arnol'd** (Ed.)
Dynamical Systems V
Theory of Bifurcations and Catastrophes
1990. Approx. 280 pp. ISBN 3-540-18173-3

Volume 6: **V. I. Arnol'd** (Ed.)
Dynamical Systems VI
Singularity Theory I
1990. Approx. 250 pp. ISBN 3-540-50583-0

Volume 16: **V. I. Arnol'd, S. P. Novikov** (Eds.)
Dynamical Systems VII
1990. Approx. 290 pp. ISBN 3-540-18176-8

Several Complex Variables

Volume 7: **A. G. Vitushkin** (Ed.)
Several Complex Variables I
Introduction to Complex Analysis
1989. VII, 248 pp. ISBN 3-540-17004-9

Volume 8: **A. G. Vitushkin, G. M. Khenkin** (Eds.)
Several Complex Variables II
Function Theory in Classical Domains. Complex Potential Theory
1991. Approx. 260 pp. ISBN 3-540-18175-X

Volume 9: **G. M. Khenkin** (Ed.)
Several Complex Variables III
Geometric Function Theory
1989. VII, 261 pp. ISBN 3-540-17005-7

Volume 10: **S. G. Gindikin, G. M. Khenkin** (Eds.)
Several Complex Variables IV
Algebraic Aspects of Complex Analysis
1990. VII, 251 pp. ISBN 3-540-18174-1

Springer-Verlag
Berlin Heidelberg New York London
Paris Tokyo Hong Kong

Springer

Encyclopaedia of Mathematical Sciences
Editor-in-chief: R.V. Gamkrelidze

Algebra

Topology

Analysis

Springer-Verlag
Berlin Heidelberg New York London
Paris Tokyo Hong Kong

Springer